"十四五"高等职业教育人工智能技术应用系列教材

RENGONG ZHINENG
TUXIANG SHIBIE YINGYONG JICHU

人工智能
图像识别应用基础

张文川 龙 翔 ◎ 主 编
年爱华 何 琳 陆益军 ◎ 副主编

中国铁道出版社有限公司
CHINA RAILWAY PUBLISHING HOUSE CO., LTD.

内容简介

本书在整体知识结构上，由浅入深地阐述了人工智能图像识别的知识体系；在实践教学上，采用目前工业应用广泛的深度学习框架，详细介绍了人工智能与图像识别、认识深度学习开发环境、机器学习和深度学习基础、数据集和预处理、图像分类、目标检测、图像分割、图像生成、深度学习模型优化、深度学习模型部署等内容。

本书在内容规划和学习方式上，采用知识要点和实践任务相结合的方式，内容与案例层层递进，全面地为读者展示从理解知识到运用知识的过程；在文字讲述和内容展示上，由点及面、图文并茂、深入浅出地阐述了人工智能图像识别领域的基本知识。

本书配套的源代码以及相关数字化资源可从中国铁道出版社有限公司网站及中育数据官网资源中心栏目下载学习。

本书适合作为高等职业院校的人工智能课程教材，也可作为普通读者学习图像识别技术应用的参考书籍。

图书在版编目(CIP)数据

人工智能图像识别应用基础/张文川,龙翔主编.—北京：
中国铁道出版社有限公司,2021.11
"十四五"高等职业教育人工智能技术应用系列教材
ISBN 978-7-113-28708-5

Ⅰ.①人… Ⅱ.①张… ②龙… Ⅲ.①人工智能-算法-应用-图像识别-高等职业教育-教材 Ⅳ.①TP391.413

中国版本图书馆 CIP 数据核字(2021)第 261957 号

书　　名	人工智能图像识别应用基础
作　　者	张文川　龙　翔

策　　划	祁　云		
责任编辑	祁　云　包　宁	编辑部电话：	(010)63549458
封面设计	尚明龙		
责任校对	孙　玫		
责任印制	樊启鹏		

出版发行：中国铁道出版社有限公司(100054,北京市西城区右安门西街8号)
网　　址：http://www.tdpress.com/51eds/
印　　刷：三河市兴达印务有限公司
版　　次：2021年11月第1版　2021年11月第1次印刷
开　　本：787 mm×1 092 mm　1/16　印张：16.75　字数：427千
书　　号：ISBN 978-7-113-28708-5
定　　价：52.00元

版权所有　侵权必究

凡购买铁道版图书,如有印制质量问题,请与本社教材图书营销部联系调换。电话：(010)63550836
打击盗版举报电话：(010)63549461

前言

人工智能从诞生起已经有60多年的历史,期间受限于计算机软硬件技术的发展,几经波折,而在最近十年间焕发生机,激发了一系列智能领域和智能产业的浪潮,并且这股浪潮必定会越发迅猛。计算机视觉是人工智能主要的发展方向之一,视觉智能计算技术在机器学习算法、大数据和图像处理计算硬件三方面的驱动下在众多领域取得了前所未有的成就,如智能监控、汽车自动驾驶、图片物体识别、智能相册、工业瑕疵检测、人脸识别、智慧医疗和智能交通等。人工智能图像识别的应用已经渗透人们生产生活的方方面面,改变了人类社会的许多生活方式,所以我们有必要从基础开始了解、学习并逐步探索这一新兴领域的应用基础。人工智能是一种可以赋能于各行各业,提升行业运行效率,降低行业运行成本的前沿技术。从应用的角度来看,人工智能图像识别的主要任务是识别和预测;从技术的角度来看,识别和预测正不断地向着更高精度、更快速度的方向进步,这正是各个行业智能化和精准化的目标。随着5G+大数据+云计算+物联网技术的革新,一个新的数字智能化时代已经到来,人工智能必将成为各行各业技术革新的基石。

本书主要特色包括:在整体知识结构上,本书由浅入深地阐述了人工智能图像识别的知识体系,适合没有接触过人工智能图像识别领域的读者全面了解现代人工智能应用技术;在实践教学上,本书采用目前工业应用广泛的TensorFlow深度学习框架,详细介绍各个项目步骤,在教学中完成各种实战案例,具有人工智能应用职业教学的价值;在内容规划和学习方式上,本书采用知识要点和实践任务相结合的方式设计,内容与案例层层递进,引导读者从理解知识到熟练运用知识;在文字讲述和内容展示上,本书由点及面、图文并茂、深入浅出地阐述人工智能图像识别领域的基本知识,力求帮助读者迅速掌握基础概念。

本书共分10个单元。单元1是人工智能与图像识别,从人工智能和图像识别导论开始,介绍了人工智能的发展历程和图像识别的主要任务;单元2是认识深度学习开发环境,从服务器硬件环境、深度学习软件框架、Python和Anaconda3环境和TensorFlow基础四方面详细阐述了深度学习开发环境的必备知识;单元3是机器学习和深度学习基础,从机器学习主要任务和算法、深度学习基础和算法等方面介绍机器学习和深度学习的算法理论基础;单元4是数据集和预处理,从数据出发,介绍了常见数据集和常见计算机任务数据集以及数据预处理的方法；单元 5～单元 8 是图像识别的核心任务，分别介绍了图像分类、目标检测、图像分割和图像生成 4 个图像识别领域内的主要任务，这部分是人工智能图像识别的核心课题，也是其他复杂图像处理任务的基石；单元 9 是深度学习模型优化，主要从模型的角度介绍了常用的模型优化方法；单元 10 是深度学习模型部署，分别从边缘端部署、浏览器前端部署和服务器部署 3 个方面详细展示了深度学习模型在不同生产环境下的部署方法和过程。

适合本书学习的对象包括：第一，打算学习并入门人工智能技术的职业院校在校学生；第二，在金融、交通、农林牧渔、制造等行业工作且希望应用人工智能解决本行业问题的工程技术人员；第三，已经对人工智能有一定的了解，想要更多、更深入地学习人工智能图像识别技术的相关人员；第四，信息和计算机科学爱好者。

本书由张文川、龙翔任主编，由年爱华、何琳、陆益军任副主编，和中育数据研发团队共同编写完成。由于编者水平有限，加之时间仓促，书中难免存在疏漏和不足之处，恳请读者批评指正。

编　者
2021 年 8 月

目 录

单元 1　人工智能与图像识别 ·· 1
　1.1　人工智能绪论 ·· 1
　　　1.1.1　人工智能的定义 ·· 1
　　　1.1.2　人工智能的发展历程 ·· 2
　　　1.1.3　人工智能的应用 ·· 4
　1.2　图像识别应用场景和主要任务 ·· 5
　　　1.2.1　图像分类 ·· 5
　　　1.2.2　目标检测 ·· 5
　　　1.2.3　图像分割 ·· 6
　　　1.2.4　图像生成 ·· 6
　小结 ··· 7
　练习 ··· 7

单元 2　认识深度学习开发环境 ·· 8
　2.1　服务器硬件环境简介 ·· 9
　2.2　深度学习软件框架简介 ·· 9
　2.3　Python 和 Anaconda3 环境简介 ·· 11
　　　2.3.1　Python 基础 ·· 11
　　　单元任务 1　使用 matplotlib 库绘图 ·· 11
　　　2.3.2　Anaconda3 集成环境 ·· 13
　　　单元任务 2　使用 Anaconda 管理开发环境 ··· 13
　　　2.3.3　Code 代码编辑器 ·· 17
　2.4　TensorFlow 基础 ··· 19
　　　2.4.1　TensorFlow 简介 ·· 19
　　　2.4.2　TensorFlow 安装 ·· 20
　　　单元任务 3　使用 conda 安装依赖库 ·· 20
　　　2.4.3　数据流图 ·· 22
　　　2.4.4　张量 ·· 22
　　　单元任务 4　使用 TensorFlow 做矩阵计算 ·· 24
　　　2.4.5　常量和变量 ·· 25

单元任务5　使用TensorFlow描述线性函数 …………………………………… 26
2.4.6　模块 ………………………………………………………………………… 28
2.4.7　高级模块 …………………………………………………………………… 29
单元任务6　使用模块和高级模块构建模型 …………………………………… 30
小结 …………………………………………………………………………………… 35
练习 …………………………………………………………………………………… 35

单元3　机器学习和深度学习基础 ……………………………………………… 36

3.1　机器学习的主要任务 …………………………………………………………… 37
3.1.1　监督学习 ……………………………………………………………………… 37
3.1.2　无监督学习 …………………………………………………………………… 37
3.1.3　分类 …………………………………………………………………………… 38
3.1.4　回归 …………………………………………………………………………… 39
3.1.5　聚类 …………………………………………………………………………… 40

3.2　机器学习算法 …………………………………………………………………… 41
3.2.1　K-近邻算法 …………………………………………………………………… 41
单元任务7　使用K-近邻识别手写数字 ………………………………………… 41
3.2.2　朴素贝叶斯 …………………………………………………………………… 43
3.2.3　线性回归 ……………………………………………………………………… 44
单元任务8　使用线性回归预测房价 …………………………………………… 44
3.2.4　支持向量机 …………………………………………………………………… 48
单元任务9　使用支持向量机实现鸢尾花分类 ………………………………… 49
3.2.5　K-均值聚类 …………………………………………………………………… 53

3.3　深度学习基础 …………………………………………………………………… 53
3.3.1　神经网络 ……………………………………………………………………… 53
单元任务10　汽车油耗预测 ……………………………………………………… 55
3.3.2　梯度下降法和批处理 ………………………………………………………… 62
3.3.3　损失函数 ……………………………………………………………………… 62

3.4　深度学习算法 …………………………………………………………………… 64
3.4.1　卷积神经网络 ………………………………………………………………… 64
单元任务11　认识卷积和池化操作 ……………………………………………… 69
3.4.2　循环神经网络 ………………………………………………………………… 70
单元任务12　循环神经网络前向传播 …………………………………………… 75
3.4.3　长短期记忆 …………………………………………………………………… 75
3.4.4　常见卷积神经主干网络 ……………………………………………………… 78

小结	86
练习	86

单元4 数据集和预处理 ································ 87

4.1 通用数据集 ································ 88
  4.1.1 MNIST 数据集 ································ 88
  4.1.2 CIFAR 数据集 ································ 88
  4.1.3 PASCAL VOC 数据集 ································ 89
  4.1.4 ImageNet 数据集 ································ 89
  4.1.5 MS COCO 数据集 ································ 89

4.2 常见计算机视觉任务数据集 ································ 91
  4.2.1 人脸数据集 ································ 91
  4.2.2 自动驾驶数据集 ································ 92
  4.2.3 医疗影像数据集 ································ 92

4.3 数据集预处理方法 ································ 93
  4.3.1 数据收集 ································ 93
  4.3.2 数据标注 ································ 94
  4.3.3 数据清洗与整理 ································ 94
  4.3.4 数据增强 ································ 95
  单元任务13 实现简单的图像数据增强 ································ 96

小结 ································ 102
练习 ································ 102

单元5 图像分类 ································ 103

5.1 图像分类问题 ································ 104
  5.1.1 图像分类概述 ································ 104
  5.1.2 图像分类类型 ································ 104
  5.1.3 图像分类步骤 ································ 105

5.2 评测指标与优化目标 ································ 106
  5.2.1 单标签分类 ································ 106
  5.2.2 多标签分类 ································ 106

5.3 图像分类的挑战 ································ 107
  单元任务14 102种花卉图像分类实战 ································ 108

小结 ································ 116
练习 ································ 116

单元 6　目标检测 ... 117

6.1　目标检测综述 ... 118
6.1.1　传统检测算法 ... 118
6.1.2　深度学习检测算法 ... 119

6.2　目标检测基础 ... 121
6.2.1　数据集 ... 121
6.2.2　评测指标 ... 122
6.2.3　损失函数 ... 123

单元任务 15　使用 Yolov3 算法实现目标检测 ... 124

单元任务 16　用 SSD 算法实现目标检测 ... 157

小结 ... 194

练习 ... 194

单元 7　图像分割 ... 195

7.1　传统图像分割方法 ... 196
7.1.1　阈值法 ... 196
7.1.2　区域生长法与超像素 ... 196
7.1.3　图切法 ... 197

7.2　深度学习图像分割 ... 198
7.2.1　基本流程 ... 198
7.2.2　反卷积 ... 198
7.2.3　多尺度与感受野 ... 199
7.2.4　图像蒙版与图像合成 ... 201

单元任务 17　使用 U-Net 模型实现城市街景图像的分割 ... 203

小结 ... 214

练习 ... 214

单元 8　图像生成 ... 215

8.1　生成对抗网络基础 ... 216

8.2　生成对抗网络结构 ... 216
8.2.1　生成网络 ... 216
8.2.2　判别网络 ... 217

8.3　生成式模型与判别式模型 ... 217

单元任务 18　代码实现 GAN 算法生成人脸图片 ... 218

小结 ... 225

练习 ... 225

单元 9　深度学习模型优化 …… 226

9.1　模型优化思路 …… 227
9.2　参数初始化 …… 227
9.3　学习率设置 …… 228
9.4　优化算法选择 …… 230
9.5　Dropout …… 231
9.6　批量归一化 …… 231
9.7　梯度爆炸/消失 …… 233

单元任务 19　优化简化版的手写体数字识别网络 …… 234

小结 …… 238

练习 …… 238

单元 10　深度学习模型部署 …… 239

10.1　模型检查点 Checkpoint …… 240
10.2　模型文件 HDF5 格式 …… 241
10.3　模型文件 SavedModel 格式 …… 241

单元任务 20　使用 TensorFlow lite 部署模型 …… 241

单元任务 21　使用 TensorFlow.js 部署模型 …… 249

单元任务 22　使用 TensorFlow serving 部署模型 …… 254

小结 …… 257

练习 …… 257

参考文献 …… 258

单元1 人工智能与图像识别

随着大数据和计算机硬件算力水平的不断发展,人工智能的各种高新前沿技术产业迅速成长并在各个领域内普遍应用。特别是在图像识别领域,机器学习和深度学习的算法模型一次次刷新分类和检测任务的评测指标,超越了人眼辨别的局限,由此基于深度学习的图像识别在全球范围内掀起了高涨的浪潮。本单元简要介绍了人工智能的定义和发展史,以及图像识别这一课题内四大主要任务,要求学生了解人工智能的基本概念,熟知并分辨图像识别的四大主要任务。

当前,人工智能图像识别在大数据处理和分析、机器学习和深度学习算法及 GPU 图形硬件计算设备的驱动下飞速发展,在各种工业领域上都逐渐应用人工智能图像识别的前沿技术。本单元就是从人工智能的定义、发展历程和工业应用介绍,要求学生了解人工智能图像识别的应用场景和主要任务。本单元的知识导图如图1.1所示。

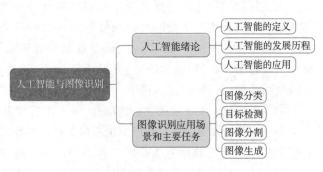

图 1.1 知识导图

课程安排

课程任务	课程目标	安排课时
认识人工智能	熟悉人工智能的发展历程和多种应用	1
熟悉图像识别的应用场景和主要任务	全面理解图像识别的四个主要任务,并通过任务的输入/输出判断图像识别的任务类型	1

1.1 人工智能绪论

1.1.1 人工智能的定义

什么是人工智能?人工智能(Artificial Intelligence,AI)最早出现于1956年美国达特茅斯学院(Dartmouth College)的讨论会上。人工智能是人们长期以来一直梦想着的可以用现代人工创造的机器设备取代人类的智能物,既包含了具体的机器设备,又涵盖了数字化的二进制程序。人工智能的定义从该名词出现到现代化人工智能的普遍应用一直被各个学派所争论。

(1) 狭义上的定义

- 1956 年,达特茅斯会议建议书:制造一台机器,该机器能模拟学习或者智能的所有方面,只要这些方面可以精确论述。
- 1975 年,人工智能专家 Minsky:人工智能是一门学科,是使机器做那些人需要通过智能来做的事情。
- 1985 年,人工智能专家 Haugeland:人工智能是计算机能够思维,使机器具有智力的新尝试。
- 1991 年,人工智能专家 Rich Knight:人工智能是研究如何让计算机做现阶段只有人才能做得好的事情。
- 1992 年,人工智能专家 Winston:人工智能是那些使知觉推理和行为成为可能的计算机系统。

(2) 广义上的定义

人工智能是现代对计算机等含有硬件和软件的具有自动化程序和解析程序的机器。传统意义上的人工智能是以大量晶体管所组成的具备大量计算能力的机器,现代的多指基于机器学习和深度学习的算法、从大数据归纳和演绎出规律的机器程序。百度百科上给出的解释是:人工智能是研究、开发用于模拟、延伸和扩展人的智能的理论、方法、技术及应用系统的一门新的技术科学。

机器学习是人工智能最主要的部分,是人工智能和模式识别领域共同研究的热点。机器学习是一种从数据中学习并归纳出有用信息的过程,让计算机不依赖确定的编码指令来自主地学习、工作。机器学习涉及概率学、统计学凸分析和算法复杂度理论等多门学科。

深度学习是由于现代机器学习中神经网络算法的大量应用所衍生出来的代名词,具体来说深度学习是机器学习的子集。深度学习主要基于数据进行特征学习,不仅是人工智能的代表性学科,也是一种建模方法。图 1.2 表示了人工智能、机器学习和深度学习的定义范畴。

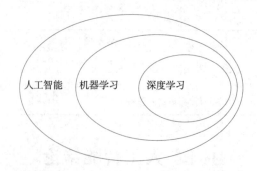

图 1.2 人工智能、机器学习和深度学习的定义范畴

1.1.2 人工智能的发展历程

人工智能的发展经历了很长时间的历史积淀。早在 1950 年,阿兰·图灵就提出了图

灵测试(Turing Test),图1.3所示为阿兰·图灵(1912—1954)。图灵测试是将人和机器放在一个小黑屋里与屋外的人对话,如果屋外的人分不清对话者是人类还是机器,那么这台机器就拥有像人一样的智能。在1956年的达特茅斯会议上,"人工智能"的概念被首次提出。在之后的十余年内,人工智能迎来了发展史上的第一个小高峰,研究者们疯狂涌入,取得了一批瞩目的成就。例如1959年,第一台工业机器人诞生;1964年,首台聊天机器人诞生。但是,由于当时计算能力的严重不足,在20世纪70年代,人工智能迎来了第一个寒冬。

早期的人工智能大多是通过固定指令来执行特定的问题,并不具备真正的学习和思考能力,问题一旦变复杂,人工智能程序就不堪重负,变得不智能了。虽然有人趁机否定人工智能的发展和价值,但是研究者们并没有因此停下前进的脚步,终于在1980年,卡内基·梅隆大学设计出了第一套专家系统——XCON。该专家系统具有一套强大的知识库和推理能力,可以模拟人类专家解决特定领域问题。从这时起,机器学习开始兴起,各种专家系统开始被人们广泛应用。不幸的是,随着专家系统的应用领域越来越广,问题也逐渐暴露出来。专家系统应用有限,且经常在常识性问题上出错,因此人工智能迎来了第二个寒冬。

图1.3 阿兰·图灵
(1912—1954)

1997年,IBM公司的"深蓝"计算机战胜了国际象棋世界冠军卡斯帕罗夫,成为人工智能史上的一个重要里程碑,从此人工智能开始了平稳向上的发展。2006年,李飞飞教授意识到了专家学者在研究算法的过程中忽视了"数据"的重要性,于是开始带头构建大型图像数据集—ImageNet,图像识别大赛由此拉开帷幕。同年,由于人工神经网络的不断发展,"深度学习"的概念被提出,深度神经网络和卷积神经网络开始不断映入人们的眼帘。2016年,Alpha GO横空出世,其与世界冠军棋手李世石进行人机大战,以4∶1的比分大获全胜,随后又依次战胜中日韩等国的多名冠军棋手,让人们感受到来自人工智能深度学习的智能技术。深度学习的发展又一次掀起人工智能的研究狂潮,这一次狂潮至今仍在持续。图1.4所示为人工智能近年来的发展历程。

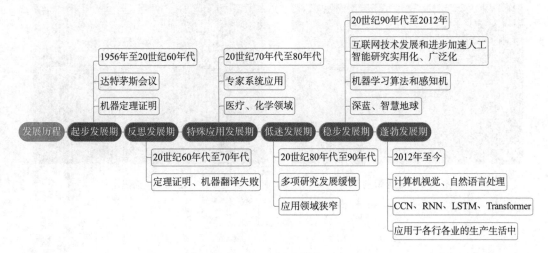

图1.4 人工智能发展历程

1.1.3 人工智能的应用

简单地说,人工智能就是让计算机或其他设备具备人脑的知识库和智能体系。例如,手机产品中各种语音辅助系统,这些辅助系统可以给人们提供更智能化的服务。人工智能主要的研究内容很多,如图1.5所示。

人工智能的应用极其广泛,从火星车到自动驾驶汽车,从信息安全到金融经济预测,从人脸扫描支付到广告商品推送,人工智能几乎遍布世界的任何一个角落。人工智能的主要应用还包括数据挖掘、计算机视觉、自然语言处理、生物特征识别、搜索引擎、医学诊断、检测信用卡欺诈、证券市场分析、DNA序列测序、语音和手写识别、战略游戏、艺术创作和机器人等,如图1.6~图1.9所示。

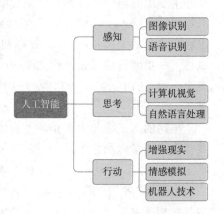

图1.5 人工智能主要研究内容

图1.6 自动驾驶

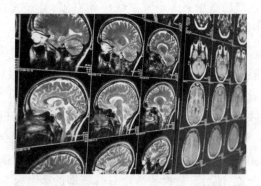

图1.7 医疗影像分析

图1.8 语音识别

图1.9 火星车

1.2 图像识别应用场景和主要任务

1.2.1 图像分类

图像分类是最为简单的一项图像识别任务,它的输入通常是含有单一类别物体的图像,输出则是图像分类的结果。如输入一组猫和狗的图片数据集,图像分类的任务就是辨别图像是猫还是狗。如图 1.10 和图 1.11 中猫狗数据集的图片。最为经典的图像分类任务数据集是 ImageNet 数据集,这个数据集由李飞飞团队维护,主要为国际 ISLVRC 比赛提供图像分类的数据集,其中训练集有 1 281 167 张图片和对应的标签,验证集有 50 000 张图片和对应的标签,测试集有 100 000 张图片。

图 1.10　狗数据集图片

图 1.11　猫数据集图片

1.2.2 目标检测

目标检测是现代人工智能图像识别相关落地项目最多的课题之一,也是目前图像识别研究的难点和热点领域。目标检测的输入可以是包含多个类别的图片和视频关键帧,不仅需要辨别前景物体的类别,而且需要确定前景物体在图像中的位置。目标检测的输出通常是边界框(Bounding Boxes),如图 1.12 和图 1.13 所示。随着深度学习和神经网络模型的计算机视觉的发展,目标检测算法也取得了突破,广泛应用于智能监控、无人驾驶和机器人感知等应用场景。比较流行的算法主要分为两类,一类是基于区域提取网络的 R-CNN 系列二阶段算法,另一类是以 Yolo、SSD、RetinaNet 等系列的一阶段算法。目前工业落地的项目上使用的算法多是一阶段算法,兼具检测效果和检测速度的优越性。

图 1.12　目标检测(一)

图 1.13　目标检测(二)

1.2.3 图像分割

图像分割的任务主要是将同一对象类别的图像部分聚类在一起,是一项像素级预测的任务,也是目前计算机视觉领域的关键问题之一。从高层次的语义角度看,图像分割是计算机视觉中场景理解的核心问题,场景理解就是程序从输入图像中推断高阶语义知识来提供完善的知识库基础。具体来说,图像分割还可以分为语义分割、实例分割、全景分割和超分辨率的子任务,广泛应用于医学图像分割、遥测图像分割和道路实景分割等应用场景。图像分割的输入通常是图像数据,输出通常为带有掩码或不同灰度的图像,输出的图像中每个像素都根据对应类别进行分类,图 1.14 所示为原始图片,图 1.15 所示为语义分割结果,图 1.16 所示为实例分割结果。

图 1.14　原始图片　　　　图 1.15　语义分割结果　　　　图 1.16　实例分割结果

1.2.4 图像生成

图像生成是计算机视觉领域近年来另一个热门话题。其中基于深度学习的 GAN 网络在图像生成上具有统治地位,其主要由生成器和判别器的相互对抗网络建模生成以假乱真的图像。图像生成任务分为有条件生成和无条件生成,输入是图像,输出是根据约束生成的图像,如图 1.17 ~ 图 1.19 所示。在计算机视觉的应用上,图像生成技术可以生成与真实数据分布一致的数据样本,弥补数据不足和数据缺失的问题。在图像修改上,图像生成技术可以实现超分辨率图像,可编辑交互式图像、风格迁移图像和图像翻译等具体任务。

图 1.17　输入图片

图1.18 真实图片

图1.19 生成图片

小　　结

本单元主要介绍人工智能的概念和图像识别的四大主要任务。从人工智能的不同层次定义和发展历程以及人工智能在现代化人类生产生活中的应用阐述了人工智能的重要性。图像识别的四大主要任务分别为图像分类、目标检测、图像分割和图像生成。本单元两项任务要求学生从不同角度认识人工智能，熟悉图像识别的应用场景和主要任务。

练　　习

练习1　简要描述人工智能以及人工智能的发展历程。

练习2　简要描述图像识别的应用场景和四大图像识别任务的具体内容。

单元2 认识深度学习开发环境

随着硬件算力水平在图形加速技术上的进步和深度学习软件框架的开源成型,人工智能在各行各业遍地开花,认识并且能够熟练使用深度学习的开发环境是入门人工智能图像识别技术应用必须掌握的技能之一。

本单元主要介绍深度学习开发环境的学习和使用。从服务器基础的硬件开发环境和软件开发框架的简介让读者建立对现代人工智能科学的基本认知,从Python和Anaconda3环境介绍了人工智能的开发语言和虚拟环境配置知识,从TensorFlow知识上零基础指导学生认识并学习使用TensorFlow深度学习框架。本单元的知识导图如图2.1所示。

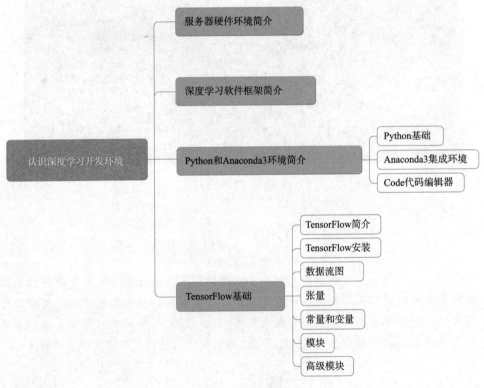

图2.1 知识导图

课程安排

课 程 任 务	课 程 目 标	安排课时
服务器硬件环境简介	熟悉服务器硬件环境和相关数据文件目录	1
深度学习软件框架简介	熟悉现代主流深度学习软件框架,支持的编程语言接口,相关优势和劣势	1
Python 和 Anaconda3 环境简介	了解并会使用 Python 语言编程,熟悉 Anaconda3 集成的环境管理系统,能熟练使用 conda 命令完成特定环境的构建,完全掌握代码编辑器 code 的使用方法	1
TensorFlow 基础	掌握 TensorFlow 1.x 的编程基础,会使用 TensorFlow 和 keras 完成简单操作图的构建	4

2.1 服务器硬件环境简介

可以使用公有云环境或本地服务器环境进行图像识别技术的学习、训练和科研。本节以后者为例,简单介绍相关硬件环境的作用。

1. 数据资源管理服务器

数据资源管理服务器上存储着大量的教学资源、数据集资源,在平台首页登录后可以通过数据集浏览服务使用相关的数据集资源或通过在线学习访问教学资源。

2. 数据处理服务器

数据处理服务器上配置有高性能的图形加速卡(GPU),在深度学习训练阶段,一般使用 GPU 作为核心的计算硬件单元。

3. 边缘推理机

边缘推理机为低功耗硬件设备,有效地模拟了手机、摄像机等移动终端。

2.2 深度学习软件框架简介

深度学习的发展离不开开源的深度学习框架,各种开源框架的社区活跃度高并且维护频繁。主流的深度学习框架有 TensorFlow、Pytorch、Caffe、CNTK 和 MXNet。这些开源框架本身质量优越,其所属的企业运作规范,迭代更新频繁,拥有庞大的社区支持和反馈,这些因素使得开源的深度学习框架得以广泛流行。

1. TensorFlow

TensorFlow 是谷歌公司推出的深度学习框架,TensorFlow 支持自动求导,用户可以方便地使用 TensorFlow 设计新的神经网络计算图和损失函数。TensorFlow 的内核由 C++ 开发,将底层细节完全抽象,避免了使用深度学习框架的开发者耗费大量时间编辑底层的 CUDA 和 C++ 代码,并且具有良好的性能保证。TensorFlow 提供了 Python、Java、C++ 和 Go 的接口。在 TensorFlow 的底层可以基于 CPU 和 GPU 进行模型的训练和推理。官网

Logo 如图 2.2 所示。

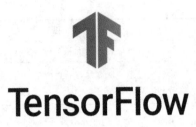

图 2.2　TensorFlow 深度学习框架

2. Pytorch

Pytorch 是基于 Torch 开发的深度学习框架,它主要由 Facebook 的人工智能小组开发。不仅能够实现 GPU 核心加速计算,同时还支持动态神经网络。Torch 是一个具有大量机器学习算法支持的科学计算框架,与 Numpy 的张量(Tensor)操作库类似,具有极高的灵活性和易用性。Torch 底层通过 C++ 开发,上层调用的接口支持 Python、C++ 和 Lua 语言。官网 Logo 如图 2.3 所示。

图 2.3　Pytorch 深度学习框架

3. Caffe

Caffe 也是一个应用广泛的老牌深度学习框架,发展时间最早,现在由 BVLC 开发和维护。Caffe 底层通过 C++ 开发,支持 Python 和 C++ 的接口。它在神经网络中的基本单元模块是 Layer。它的接口简洁,容易上手,训练速度快,底层通过 cuDNN 加速库训练网络模型。Facebook 人工智能实验室与应用机器学习团队开发的 Caffe2 在 Caffe 的基础上做了很多优化,训练和推理的运行速度提升了,并且具有了跨平台、可扩展的优点。官网 Logo 如图 2.4 所示。

图 2.4　Caffe 深度学习框架

4. CNTK

CNTK 是微软开源的深度学习框架,主要被用在 Cortana 数字助理和 Skype 翻译应用中的语音识别框架。CNTK 支持多个 GPU 或服务器并行计算,其使用 C++ 作为底层开发,支持 Python 和 C++ 的模块接口。官网 Logo 如图 2.5 所示。

图 2.5　CNTK 深度学习框架

5. MXNet

MXNet 是 DMLC 推出的开源深度学习框架,是亚马逊主推的深度学习框架。AWS 官方服务器都是使用 MXNet 进行深度学习编程和部署的。MXNet 上层接口支持 C++、Python、R、Julia 和 Go 等语言,由于其命令式和符号式的编程使其应用广泛。MXNet 拥有类似 Theano 和 TensorFlow 的数据流图,为多 GPU 架构提供了良好的配置,其支持分布式的优越性使其被广泛推广。官网 Logo 如图 2.6 所示。

图 2.6　MXNet 深度学习框架

除上述框架外,深度学习框架还有 Theano、PaddlePaddle、MegEngine 等,这些框架有的已经停止维护,有的还在不断地进行更新迭代。这里推荐学习使用其中一种较为流行的框架后,不断在项目实践过程中进行代码练习,逐渐达到工程应用开发的水准。

2.3　Python 和 Anaconda3 环境简介

2.3.1　Python 基础

Python 是一种面向对象的解释型动态计算机程序设计语言,在人工智能、科学计算、数据科学、Web 开发、系统运维、大数据及云计算领域有着数量庞大且功能完善的第三方库,能够实现不同领域业务的应用开发。因为深度学习框架的使用和数据处理的优势所在,Python 已经成为人工智能算法指定的开发语言。在 Linux 平台下,部分发行版(如 ubuntu 20.4 系统)已经集成了 Python 3.8 版本的解释器。Python 解释器配置有相应的 pip 包管理器,可以使用命令从官方源下载所需的第三方库。

单元任务 1　使用 matplotlib 库绘图

使用 pip 安装 matplotlib 库,并使用 matplotlib 绘制 $y=x^2$ 曲线图。

matplotlib 是 Python 的绘图库,它可与 NumPy 一起使用,提供了一种轻量级的 MatLab 开源高效替代方案。它也可以和图形工具包 PyQt 等工具一起配合使用,能够完成日常科学计算中多种数学图库可视化任务。

步骤 1:使用 pip 安装 matplotlib 库。

① 使用 Code 代码编辑器打开数据处理服务器的控制台终端,如图 2.7 所示。

图 2.7 控制台终端

② 从终端输入 pip list 命令查看当前系统环境的 Python 第三方软件包,如图 2.8 所示。

图 2.8 系统环境的 Python 第三方软件包

③ 从终端输入 pip install matplotlib 命令,安装 matplotlib 库,如图 2.9 所示。

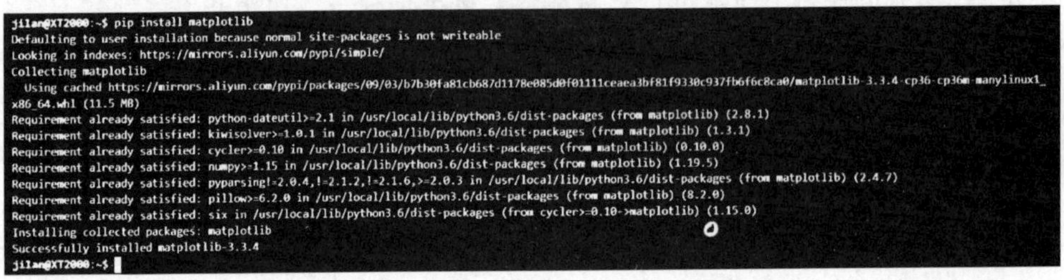

图 2.9 安装 matplotlib 库

④ 再次使用 pip list 命令查看当前系统环境的 Python 第三方软件包,可以找到系统环境下安装的 matplotlib 库,如图 2.10 所示。

图 2.10 系统环境下的 matplotlib 库

步骤 2:使用 matplotlib 绘制 $y = x^2$ 曲线图。

从控制台终端创建文件 task1.py。

```
touch task1.py
```

【代码清单 task1.py】

```
import numpy as np
import matplotlib.pyplot as plt

x = np.arange(-3,3,0.1)
y = x**2
fig = plt.figure()
plt.plot(x,y)
plt.title('y=x**2')
plt.xlabel('x')
plt.ylabel('y')
plt.savefig('task1.jpg')
print('Done.')
```

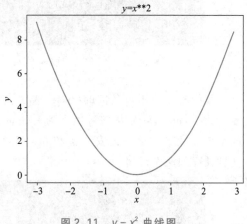

生成 task1.jpg 图像，如图 2.11 所示。

图 2.11　$y = x^2$ 曲线图

2.3.2　Anaconda3 集成环境

Anaconda3 是一个用于环境和包管理的集成 Python 发行版的工具，支持跨平台，能够让用户在数据科学中轻松地管理深度学习集成开发环境，并且也配置了专门的包管理器，让用户安装常用程序包更为便捷。Anaconda3 的核心为 conda 工具，conda 工具的命令都是以 conda 开头的，其核心功能是包管理和环境管理。Anaconda3 官网 Logo 如图 2.12 所示。

1. 包管理

不同的第三方库在安装和使用过程中都会存在版本匹配和兼容性问题，在实际工程中经常会使用大量的第三方安装包，若手动进行匹配则耗时耗力，因此自动化包管理的方案在集成开发环境中显得非常重要。

2. 环境管理

用户可以使用 conda 创建多个虚拟环境，不同的工程运行在相应的虚拟环境下，各个虚拟环境都是独立的、互不影响的开发运行环境，这样的多种环境并存的集成开发环境可以方便地解决多个版本的 Python 和不同深度学习框架的平行开发，达到任意切换的效果。

图 2.12　Anaconda3 开发环境

单元任务 2　使用 Anaconda 管理开发环境

使用 Anaconda3 新建名称为 tf1.15 的虚拟环境，指定 Python 解释器版本为 3.6，并在环境中安装 numpy 科学计算库，完成半径为 9 的圆面积计算。

Anaconda3 原生自带有 base 环境，注意这里的 base 环境与操作系统环境是两个不同的环境。在命令行中输入 conda activate base 命令，从系统环境切换至 base 环境，如图 2.13 所示。

图 2.13 激活 base 环境

其中命令行用户名前的(base)表示当前处于 base 环境中,可以使用包管理器命令 pip list 或者 conda list 显示该环境下的 Python 安装包,如图 2.14 所示。如果仔细观察可以发现系统环境下的第三方安装包和 base 环境下的第三方安装包有所不同,因为这是完全独立且共存的两个环境。

图 2.14 base 环境的安装包

步骤 1:使用 conda 新建名称为 tf1.15 的虚拟环境,并指定 Python 解释器版本为 3.6。命令行输入如下命令,创建名称为 tf1.15 的虚拟环境,如图 2.15 所示。

```
conda create -n tf1.15 python=3.6
```

图 2.15 conda 创建虚拟环境

命令行输入 y 后确认下载基础的标准库,如图 2.16 所示。

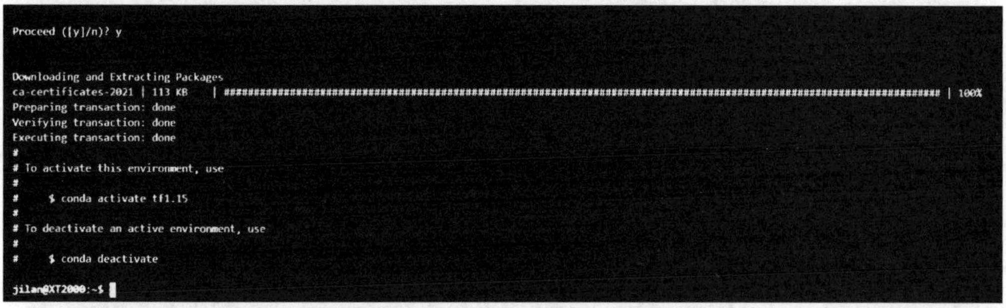

图 2.16　安装新环境标准库

命令行输入 conda activate tf1.15 进入 tf1.15 虚拟环境。命令行继续输入 conda list 显示该环境下已安装的软件包,如图 2.17 所示。

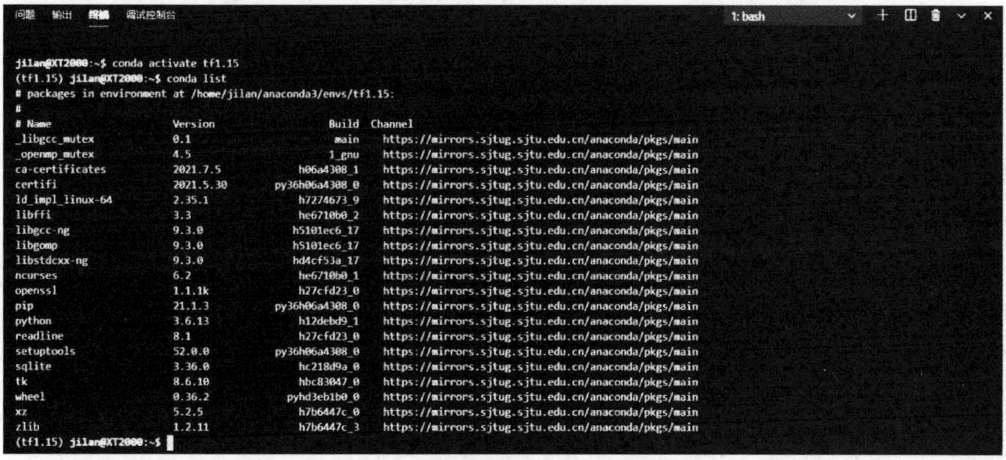

图 2.17　激活虚拟环境并显示安装包列表

命令行输入 python- -version 查看当前环境下 Python 的版本,命令行输入 conda deactivate 退出当前虚拟环境,如图 2.18 所示。

图 2.18　查看 Python 版本并退出环境

步骤 2:在 tf1.15 环境中安装 numpy 科学计算库。

命令行输入 conda activate tf1.15 再次进入 tf1.15 虚拟环境。命令行输入 pip install numpy 安装 numpy 科学计算库,也可以输入 conda install numpy 命令进行安装,如图 2.19 所示。

图 2.19　在新环境中安装 numpy 科学计算库

命令行输入 python 打开交互窗口,如图 2.20 所示。

图 2.20　Python 交互窗口

命令行输入代码测试 numpy 科学计算库是否安装成功,如图 2.21 所示。

```
import numpy as np
```

图 2.21　验证 numpy 库的安装

步骤 3:计算半径为 9 的圆面积。
命令行逐行输入命令,计算结果如图 2.22 所示。

```
import numpy as np
r = 9
s = np.pi * r ** 2
print(s)
print('%.2f' % s)
```

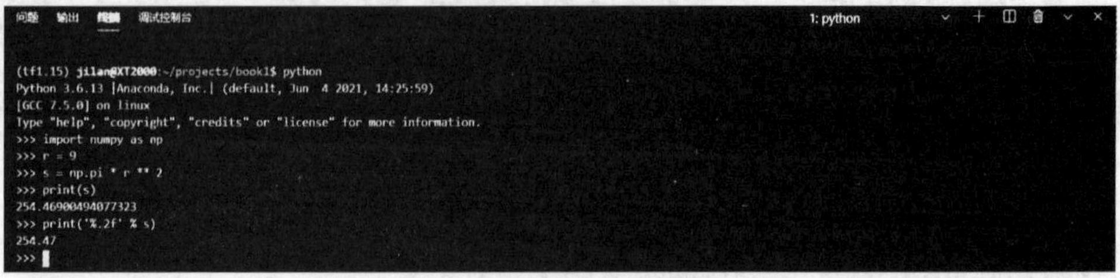

图 2.22　圆面积计算

命令行输入 conda remove -n tf1.15 - -all 可删除 tf1.15 的虚拟环境。

2.3.3 Code 代码编辑器

Code 代码编辑器主页面有三个重要的区域,分别是工程目录区、代码编辑区和控制台终端区,如图 2.23 所示。工程目录区显示打开的工程目录,代码编辑区用于编写代码,控制台终端区用于输入命令行。

图 2.23 Code 代码编辑器工作区

在 Code 扩展库中可以安装各种编辑器的插件,下面在应用商店中搜索扩展 Python,如图 2.24 所示,安装第一个 ms-python 插件后可以使 Code 编辑器具有更好的编程体验。

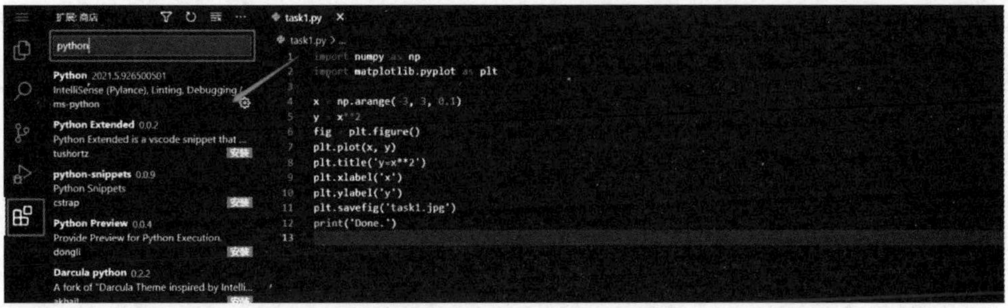

图 2.24 插件安装

Code 编辑器底端是状态栏,如图 2.25 所示,其中显示 Code 编辑器的服务地址、环境 Python 解释器、错误和警告数、编辑区代码行数和列数、缩进格式、编码格式、换行模式、语言模式等信息。

图 2.25 Code 编辑器状态栏

通过单击 Python 插件提供的环境 Python 解释器可以更换其他环境 Python 解释器,如

图 2.26 所示。

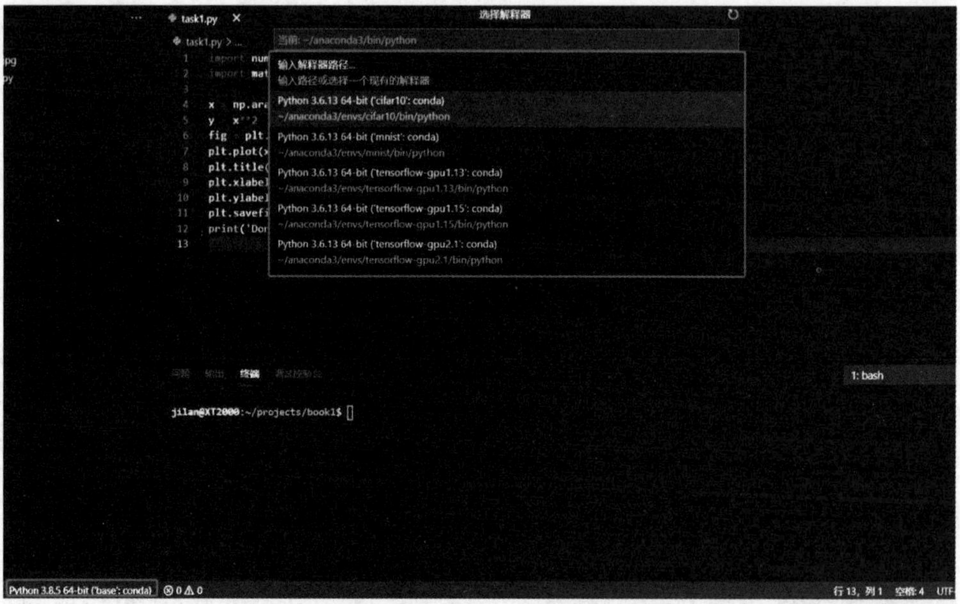

图 2.26　切换不同 Python 解释器

在某一虚拟环境下也可以创建具有良好交互页面的 Jupyter Notebook 编辑器。例如，在之前创建的 tf1.15 环境下开启 Jupyter Notebook 编辑器，其编辑器以 tf1.15 环境下的解释器作为运行 Python 代码的内核。在 tf1.15 环境下，打开 Code 编辑器的终端命令行，输入 pip install notebook 命令，完成后再从终端命令行，输入 jupyter notebook 命令，启动页面服务，通过浏览器打开服务的网址即可进入 Jupyter Notebook 编辑器，如图 2.27 所示。

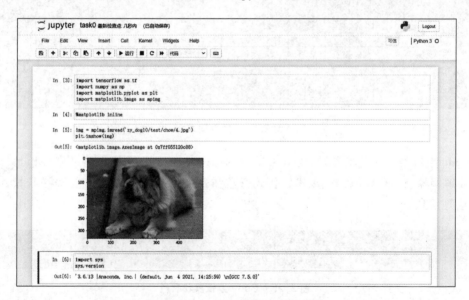

图 2.27　Jupyter Notebook 编辑器

以上 Code 代码编辑器和 Jupyter Notebook 编辑器是需要掌握的基本技能,读者可选择其一,不断地熟悉编辑器的各项功能,逐步达到灵活使用的水平。在后续的代码编程实践中可以学习到更多的编辑技巧和编程技能,只有不断练习、日积月累,才能取得突破和进步。

2.4　TensorFlow 基础

2.4.1　TensorFlow 简介

TensorFlow 是目前主流深度学习框架之一,其库中几乎包含了所有机器学习和深度学习相关的辅助函数和封装类,官方网站如图 2.28 所示,在其框架下做各种神经网络算法的开发可以极大减轻工作量,在入门阶段可以不需要深入理解相关优化算法、分布式的底层细节也可以完成对于深度学习神经网络的搭建、训练、评估、测试和部署步骤。TensorFlow 最初是由谷歌脑研究组的研究员和工程师们开发出来的,主要用于进行机器学习和深度神经网络方面的研究,后来逐渐发展成为广泛通用的深度学习主流框架之一。

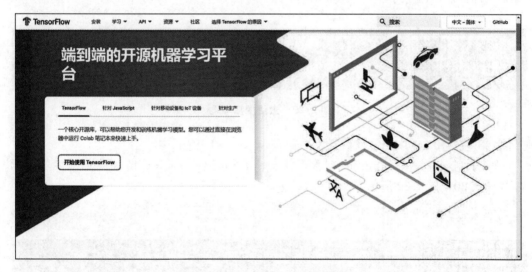

图 2.28　TensorFlow 官方网站

TensorFlow 主要采用数据流图(Data flow graphs)规划计算流程、进行数值计算,图中的节点(Node)表示数学操作;图中的线(Edges)表示在节点之间相互传递信息的多维数组;记录多维数组信息量的数据为张量(Tensor)。TensorFlow 的发行版本主要分为 1.x 和 2.x,在 1.9 之后的发行版本都支持动态图和静态图的构建,在 2.x 版本中 TensorFlow 将支持动态图的 eager 模式设置为默认的执行模式,建议学习动态图模式编程,但是最好也具备较为复杂的静态图模式编程的能力。TensorFlow 提供了 Python、C++和 JavaScript 等易于快速开发编程语言的接口,并且还配置有 TensorBoard 可视化工具。TensorFlow 具有灵活的架构,支持在多种平台上进行计算,如 CPU、GPU、TPU、云服务器和移动设备,具有灵活、快速和适用性广泛的优越性。

2.4.2　TensorFlow 安装

TensorFlow 支持多数平台的系统环境,有多个发行版本。根据运行的硬件库的不同可以划分为 CPU 版本和 GPU 版本,CPU 版本适合轻量级数据的训练推理测试处理,通常耗时较长、处理速度较慢,GPU 版本适合普遍的大数据量级的训练推理测试处理,一般耗时较短、处理速度较快。数据处理服务器平台搭载有高性能的 GPU 图形加速设备,因此推荐使用 GPU 版本的 TensorFlow 进行计算。下面在虚拟环境中安装特定版本的 TensorFlow。

单元任务 3　使用 conda 安装依赖库

使用 conda 在 tf1.15 环境中安装 GPU 版本、发行版本号为 1.15 的 TensorFlow 计算库。

步骤 1:进入 tf1.15 开发环境。

命令行输入命令进入 tf1.15 开发环境,如图 2.29 所示。

```
conda activate tf1.15
```

图 2.29　激活虚拟环境

步骤 2:使用 conda 包管理器安装 tensorflow-gpu=1.15 版本。

命令行输入命令,使用 conda 包管理器安装 tensorflow-gpu=1.15 版本的深度学习框架,如图 2.30~图 2.32 所示。

```
conda install tensorflow-gpu=1.15
```

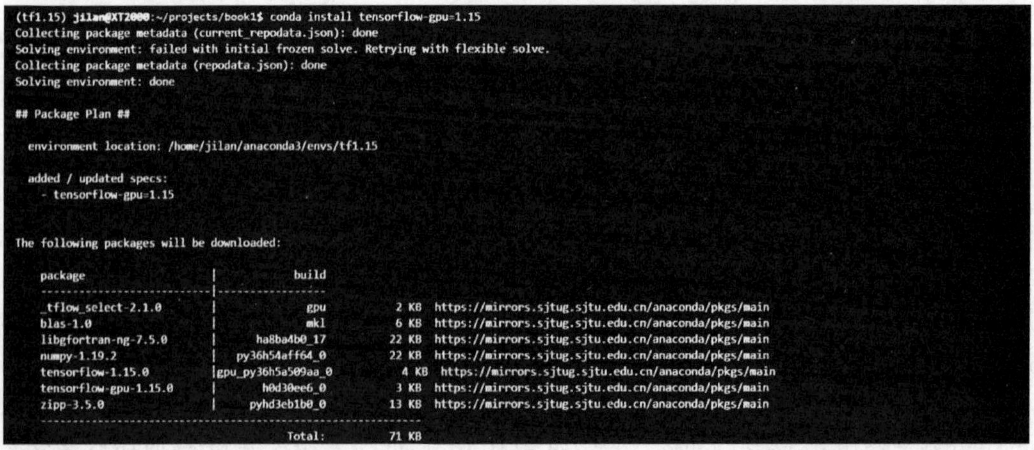

图 2.30　安装 TensorFlow 深度学习框架(一)

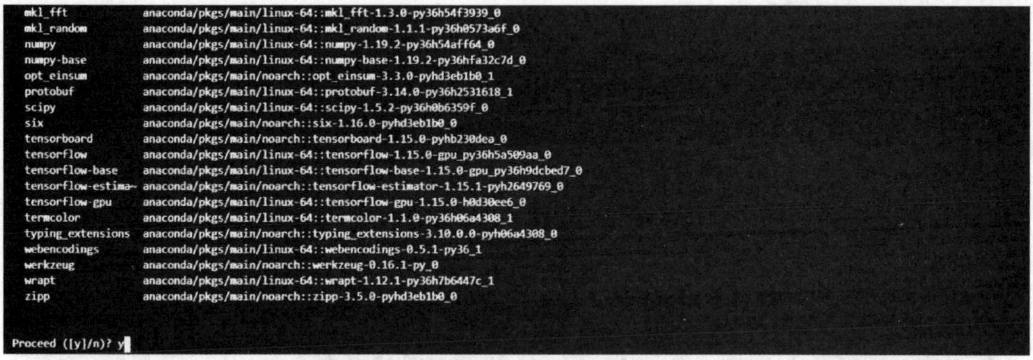

图 2.31 安装 TensorFlow 深度学习框架(二)

输入 y 确认。

图 2.32 安装 TensorFlow 深度学习框架(三)

命令行输入命令查看当前 tf1.15 环境下的安装包列表,可以看到多个 TensorFlow 的库名称,如图 2.33 所示。

```
conda list
```

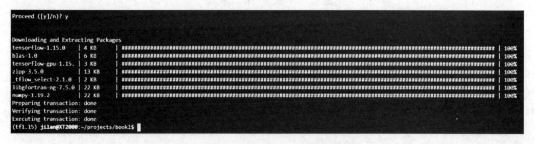

图 2.33 tf1.15 环境的安装包

步骤 3:验证安装的 GPU 版本的 TensorFlow 是否可用。

命令行输入 python,启动交互对话框,如图 2.34 所示。

命令行逐行输入命令,验证安装的 GPU 版本的 TensorFlow 是否可用,如图 2.35 和图 2.36 所示。

```
import tensorflow as tf
tf.test.is_gpu_available()
```

图 2.34　启动 Python 交互命令行

图 2.35　验证 GPU 驱动

图 2.36　验证 GPU 驱动结果

在加载一连串 CUDA 库后，输出为 True 表示当前环境下的 tensorflow-gpu 可用。

2.4.3　数据流图

数据流图是使用节点(Node)和有向线(Edge)来描述的数学计算，又称计算图。在用 TensorFlow 进行科学计算时，通常先创建一个计算图，然后将数据载入计算图中进行数据计算，如图 2.37 所示。

1. 节点

计算图中一般用圆圈、椭圆或方框表示，计算图在 TensorBoard 可视化中可以形象化地显示出来。节点通常用来表示执行的数学操作，数据输入的起点和数据输出的终点也可以表示为节点。

2. 线

计算图中一般用箭头表示线，代表节点与节点之间的信息输入、输出和传递关系，其中传递的信息就是可变维度的张量。

2.4.4　张量

张量是计算图中节点之间相互传递数据的表现形式。一维数组、二维数组和 N 维数组等都可以看作张量，表示操作的输出量。tf.Tensor(op,value_jndex,dtype)参数描述见表 2.1，张量的属性见表 2.2，TensorFlow 中张量常见数据类型见表 2.3。

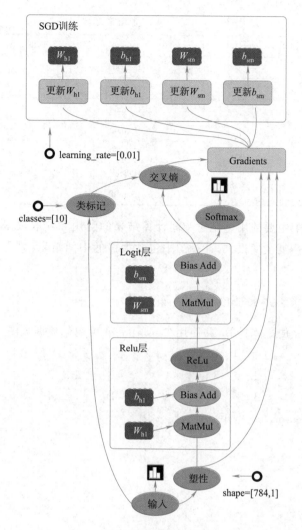

图 2.37 数据流图

表 2.1 tf.Tensor(op,value_index,dtype)参数描述

参数	描述
op	Tensor 操作节点
value_index	生成 Tensor 的节点索引
dtype	Tensor 数据类型

表 2.2 张量的属性

属性	描述
Tensor.device	计算 Tensor 的硬件
Tensor.dtype	Tensor 数据类型
Tensor.graph	包含 Tensor 的计算图
Tensor.name	Tensor 命名空间
Tensor.op	Tensor 操作节点
Tensor.shape	Tensor 形状

表 2.3　TensorFlow 中张量常见数据类型

数 据 类 型	Python API
32 位浮点数	tf.float32
64 位浮点数	tf.float64
64 位有符号整型	tf.int64
32 位有符号整型	tf.int32
可变长字节数组	tf.string
布尔型	tf.bool
8 位无符号整型	tf.uint8

张量在计算图中主要的用途是对中间计算结果的引用和获取数据流图计算结果。在构建深层网络时计算复杂度很大，计算图中包含大量的中间结果，只用张量可以极大地提升代码的可读性。

单元任务 4　使用 TensorFlow 做矩阵计算

用 TensorFlow 完成矩阵计算，并使用 TensorBoard 可视化数据流图。

$$Y = \begin{bmatrix} 1 & 4 & 7 \\ 2 & 5 & 8 \\ 3 & 6 & 9 \end{bmatrix} * \begin{bmatrix} 2 & 1 & 2 \\ -1 & 1 & 4 \\ 3 & 0 & -1 \end{bmatrix} + \begin{bmatrix} 1 & -1 & 0 \\ 2 & 2 & 2 \\ 1 & 4 & 5 \end{bmatrix}$$

步骤 1：使用 TensorFlow 做矩阵计算。

命令行输入：

```
touch task4.py
```

【代码清单 task4.py】

```python
import tensorflow as tf

A = tf.constant([[1,4,7],[2,5,8],[3,6,9]],name = 'matrix_A')
B = tf.constant([[2,1,2],[-1,1,4],[3,0,-1]],name = 'matrix_B')
C = tf.constant([[1,-1,0],[2,2,2],[1,4,5]],name = 'matrix_C')

node_matmul = tf.matmul(A,B,name = 'node_matmul')
node_add = tf.add_n([node_matmul,C],name = 'node_add')

with tf.Session() as sess:
    writer = tf.summary.FileWriter("./log_4",sess.graph)
    Y = sess.run(node_add)
    print(Y)
    writer.close()
```

步骤 2：运行程序。

命令行输入命令，运行结果如图 2.38 所示。

```
python task4.py
```

步骤 3：使用 TensorBoard 可视化矩阵计算数据流图。

图 2.38 矩阵计算

使用 TensorBoard 显示矩阵计算的数据流图，命令行输入：

```
#注意--host 参数值设置为数据处理服务器的 IP 地址
tensorboard --host 172.16.33.106 --port 8888 --logdir log_4
```

在浏览器新的标签页输入网址 http://172.16.33.106:8888/，即可访问 TensorBoard 页面，如图 2.39 所示。

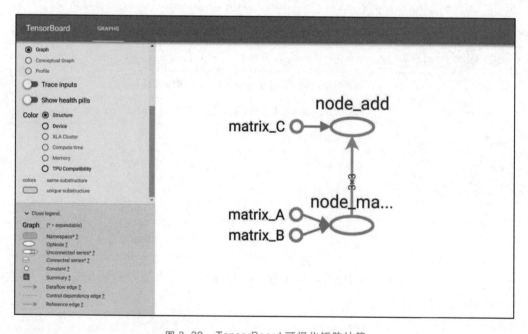

图 2.39 TensorBoard 可视化矩阵计算

2.4.5 常量和变量

1. 常量

常量是静态的数据，赋值后就不随时间发生变化，TensorFlow 常量用 tf.constant( ) 表示，见表 2.4。

表 2.4 tf.constant(value,dtype=None,shape=None,name='Const')参数描述

参　　数	描　　述
value	value 可以为 np.ndarray,也可以是列表
dtype	常量数据类型
shape	常量形状
name	常量命名空间标识符

2. 变量

变量是动态的数据,可以随着时间更新保存的数值,TensorFlow 变量使用 tf.Variable()类构建。使用 tf.Variable()方法创建变量需要输入初始值,初始值的形状和类型决定了变量的形状和类型,见表 2.5。

表 2.5 tf.Variable(initial_value=None,trainable=None,validate_shape=True,name=None,dtype=None,expected_shape=None,import_scope=None,shape=None)参数描述

参　　数	描　　述
initial_value	初始化变量的 Tensor 值
trainable	如果为 true,则变量值可变;如果为 False,则变量值不可变
validate_shape	如果为 true,则初始化变量值唯一;如果为 False,则初始化变量值不唯一
name	变量命名空间标识符
dtype	变量数据类型
expected_shape	广播的变量形状
import_scope	变量命名空间
shape	变量形状

注意:在 TensorFlow 1.x 中,所有变量在计算图中运行时都要进行初始化操作,而在 TensorFlow 2.x 中则不用。

```
with tf.Session()as sess:
init=tf.global_variables_initializer()
sess.run(init)
```

单元任务 5　使用 TensorFlow 描述线性函数

使用 TensorFlow 描述线性函数 $Y = WX + b$,其中 W 为随机张量,维度为(64,64),输入 X 的维度为(64,1),b 为全 1 向量,维度为(64,1),并使用 TensorBoard 可视化数据流图。

步骤 1:使用 TensorFlow 描述线性函数 $Y = WX + b$。

新建 task5.py 文件,在 Code 编辑器中输入如下代码。

【代码清单 task5.py】

```
import numpy as np
import tensorflow as tf

with tf.variable_scope('linear_function'):
    X = tf.placeholder(tf.float32,[64,1],name = 'X')
    W = tf.Variable(tf.random_normal([64,64],stddev = 1,seed = 1),name = 'W')
    b = tf.Variable(tf.ones([64,1]),name = 'b')
    Y = tf.matmul(W,X) + b
init = tf.global_variables_initializer()
with tf.Session() as sess:
    writer = tf.summary.FileWriter("./log_5",sess.graph)
    sess.run(init)
    rand_array = np.random.rand(64,1)
    print(sess.run(Y,feed_dict = {X: rand_array}))
    writer.close()
```

步骤2：运行 task5.py 程序。

命令行输入命令，运行结果如图2.40所示。

```
python task5.py
```

图 2.40　运行结果

步骤3：使用 TensorBoard 可视化线性函数 Y = WX + b 数据流图。

命令行输入：

```
tensorboard --host 172.16.33.106 --port 8888 --logdir log_5
```

在浏览器新的标签页输入网址 http://172.16.33.106:8888/，访问 TensorBoard 页面，如图2.41所示，可以形象地看到为线性函数编写的计算图，图中有多个节点和边，整个操作都处于命名空间 linear_function 中。

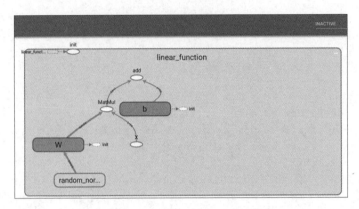

图2.41　TensorBoard可视化线性函数

2.4.6　模块

TensorFlow框架内构建了很多高层次的API,可以显著减少编写程序的代码量,其中包含众多网络结构相关函数和数据载入、数据处理的方法。

1. tf.data.Dataset

tf.data.Dataset是TensorFlow内置的数据输入模块,提供了专门用于数据输入的多种方法,可以高效地实现数据载入、数据增强和数据随机乱序等功能。例如,最简单的数据载入方式是从列表载入张量数据。

在虚拟环境tf1.15的命令行输入python,打开交互命令行,使用tf.enable_eager_execution()可以开启TensorFlow 1.x版本的动态图模式,使用tf.data.Dataset.from_tensor_slices方法将列表[1,2,3]按第一个维度转换为张量Tensor,如图2.42所示。

```
>>>import tensorflow as tf
>>>tf.enable_eager_execution()
>>>dataset=tf.data.Dataset.from_tensor_slices([1,2,3])
>>>for element in dataset:
...    print(element)
tf.Tensor(1,shape=(),dtype=int32)
tf.Tensor(2,shape=(),dtype=int32)
tf.Tensor(3,shape=(),dtype=int32)
>>>exit()
```

图2.42　列表数据转换为张量数据

2. tf.layers

tf.layers是TensorFlow 1.x版本内置的构建神经网络的模块,在TensorFlow 2.x中被移除,其中封装了很多底层的函数和基本的神经网络结构,在熟悉TensorFlow底层后可以直接使用tf.layers提供的高级API实现各种复杂的建模任务,能够省去大量的代码。

3. tf.nn

tf.nn 相对于 tf.layers 模块更为底层一些,tf.nn 模块中提供了大量激活函数和神经网络层参数计算方法。通常使用较多的函数有 tf.nn.sigmoid()、tf.nn.softmax()、tf.nn.relu()、tf.nn.tanh()、tf.leaky_relu()等。

2.4.7 高级模块

1. Keras

Keras 库是最常用的 TensorFlow 高级核心 API,隐藏了数据流和底层结构的很多细节,其库中具有大量可直接使用的神经网络结构和常用模块。Keras 的代码完全由 Python 编写,在使用 TensorFlow 作为其后端时,较好地兼容了 TensorFlow 底层的各种库函数和核心模块。在 TensorFlow 2.x 发行版本中集成了 Keras 模块。对于常见的神经网络层,Keras 均实现了完美的封装,简单易用,特别适合初学者构建深度学习模型。

(1)导入库

```
from tensorflow.keras import Sequential
from tensorflow.keras.layers import Dense,Dropout,Flatten,Conv2D,MaxPooling2D
#对于 TensorFlow<1.15 版本
#from keras.models import Sequential
#from keras.layers import Dense,Dropout,Flatten,Conv2D,MaxPooling2D
```

(2)序列构建神经网络模型

```
#构建序列
model = Sequential()
#序列加入卷积层
model.add(Conv2D(...))
#序列加入池化层
model.add(MaxPooling2D(...))
#序列加入全连接层
model.add(Dense(...))
#序列加入随机失活
model.add(Dropout(...))
```

2. TensorLayer

TensorLayer 库也是一个基于 TensorFlow 的高层次的 API 库,灵活性很强,具有属于自己的一套语法,可以简单地实现动态网络结构,运行计算速度较快。相比于 Keras,TensorLayer 高级模块使用的开发人数并不多,不适合初学时使用。

TensorLayer 库安装:

```
pip install tensorlayer
```

3. TFLearn

TFLearn 库同样具有高层次的 API 设计,代码可读性和灵活性与 Keras 库相当,其中屏蔽了大量 TensorFlow 底层难以理解的东西,同样也封装了大量常用的神经网络结构。相比 Keras,TFLearn 对初学者并不友好,适合有一定编程基础的开发者使用。

TFLearn 库安装:

```
pip install tflearn
```

对于这些高级模块,推荐熟悉掌握其中一个即可,这里建议读者熟练掌握 TensorFlow 内置的 Kears 高级模块。Keras 高级模块不仅适用于各种深度学习模型的构建,而且在封装和使用上简单易学,受到越来越多的开发者青睐。

单元任务6 使用模块和高级模块构建模型

使用 Keras 高级模块构建简单的神经网络模型,完成 zy_class10 图像数据的分类,输入分辨率为 96×96×3,输出为 10 个类别。

步骤1:创建新的工程目录并准备数据集,这里运行环境为 tf2.1。

```
#创建工程目录 zy_class10
mkdir zy_class10
```

zy_class10 数据集为 10 个类别的常见物体分类数据集,数据集所在位置为数据资源服务器。下载所需数据集并解压到 dataset 文件目录。

命令行输入:

```
wget http://172.16.33.72/dataset/zy_class10.tar.gz
tar-zxvf zy_class10.tar.gz
mv zy_class10 dataset
rm zy_class10.tar.gz
```

查看数据集文件夹,命令行输入:

```
tree dataset -L 2
```

显示 dataset 数据集。dataset 下包含训练集和测试集两个目录,其中包括 10 类分辨率为 96×96×3 的常见物体图片,标签类别即为文件名。其中 10 种常见类别有 airplane(飞机)、bird(鸟)、car(汽车)、cat(猫)、deer(鹿)、dog(狗)、horse(马)、monkey(猴)、ship(船)和 truck(卡车)。

```
zy_class10
├──test
│   ├──airplane
│   ├──bird
│   ├──car
│   ├──cat
│   ├──deer
│   ├──dog
│   ├──horse
│   ├──monkey
│   ├──ship
│   └──truck
└──train
    ├──airplane
    ├──bird
    ├──car
    ├──cat
    ├──deer
    ├──dog
    ├──horse
    ├──monkey
    ├──ship
    └──truck
22 directories,0 files
```

步骤 2：使用 tf.data.Dataset 载入数据集。

载入 10 个常见类别的图片地址，输出全部目录下的图片地址

```
def all_images_path(path):
    #返回所有图片文件地址
    all_images_path = glob.glob(os.path.join(path,'*/*.jpg'))
    random.shuffle(all_images_path)
    return all_images_path
```

载入 10 类图片的标签：标签对应为名称和索引值。

```
def label_map():
    #类标签函数
    name0 = ['airplane','bird','car','cat','deer','dog','horse','monkey',
            'ship','truck']
    name1 = ['飞机','鸟','汽车','猫','鹿','狗','马','猴','船','卡车']
    label_en_index = dict((name0,index) for index,name0 in enumerate(name0))
    label_zh_index = dict((name1,index) for index,name1 in enumerate(name1))
    index_en_label = dict((index,name) for name,index in label_en_index.items())
    index_zh_label = dict((index,name) for name,index in label_zh_index.items())
    return (label_en_index,label_zh_index,index_en_label,index_zh_label)
```

载入图片数据的标签映射：

```
def load_label(all_path):
    #载入图片数据,输入图片地址,返回 label 映射
    all_label = list()
    label_en_index = label_map()[0]
    for path in all_path:
        label_name = os.path.basename(os.path.dirname(path))
        label = label_en_index[label_name]
        all_label.append(label)
    return all_label
```

载入图片数据，输入一个图片地址，返回预处理的 Tensor：

```
def load_image(image_path):
    #载入图片数据,输入一个图片地址,返回预处理的 Tensor
    image = tf.io.read_file(image_path)
    image = tf.image.decode_jpeg(image,channels=3)
    image = tf.image.resize(image,SIZE)
    image = tf.cast(image,tf.float32)/255
    return image
```

使用 tf.data.Dataset.from_tensor_slices() 载入数据集，其中 tf.data.Dataset.from_tensor_slices() 将图片和标签载入为数据队列；map 方法可转换图片地址为 Tensor 数据；zip 方法将图片和对应的标签合并为元组整体；repeat 方法可重复选取数据队列；shuffle 方法指定了数据队列中元素乱序；batch 方法指定了数据队列的批处理数据量；返回值 train_ds、test_ds 分别为训练集数据和测试集数据。

```python
def load_dataset():
    #使用 tf.data.Dataset.from_tensor_slices()载入数据集
    train_images_path = all_images_path(TRAIN_PATH)
    train_label = load_label(train_images_path)
    test_images_path = all_images_path(TEST_PATH)
    test_label = load_label(test_images_path)

    train_image_ds = tf.data.Dataset.from_tensor_slices(train_images_path)
    train_image_ds = train_image_ds.map(load_image,num_parallel_calls=AUTOTUNE)
    train_label_ds = tf.data.Dataset.from_tensor_slices(train_label)
    train_ds = tf.data.Dataset.zip((train_image_ds,train_label_ds))
    train_ds = train_ds.repeat().shuffle(buffer_size=1000).batch(BATCH_SIZE)

    test_image_ds = tf.data.Dataset.from_tensor_slices(test_images_path)
    test_image_ds = test_image_ds.map(load_image,num_parallel_calls=AUTOTUNE)
    test_label_ds = tf.data.Dataset.from_tensor_slices(test_label)
    test_ds = tf.data.Dataset.zip((test_image_ds,test_label_ds))
    test_ds = test_ds.batch(BATCH_SIZE)
    return train_ds,test_ds
```

步骤3：使用 tf.keras 高级模块构建一个简单的神经网络模型。

```python
def build_model():
    #定义神经网模型
    model = Sequential()
    model.add(Conv2D(32,(3,3),activation='relu',input_shape=(96,96,3)))
    model.add(BatchNormalization())
    model.add(Conv2D(32,(3,3),activation='relu'))
    model.add(BatchNormalization())
    model.add(MaxPooling2D(pool_size=(2,2)))
    model.add(Dropout(0.25))

    model.add(Conv2D(64,(3,3),activation='relu'))
    model.add(BatchNormalization())
    model.add(Conv2D(64,(3,3),activation='relu'))
    model.add(BatchNormalization())
    model.add(MaxPooling2D(pool_size=(2,2)))
    model.add(Dropout(0.25))

    model.add(Conv2D(128,(3,3),activation='relu'))
    model.add(BatchNormalization())
    model.add(Conv2D(128,(3,3),activation='relu'))
    model.add(BatchNormalization())
    model.add(MaxPooling2D(pool_size=(2,2)))
    model.add(Dropout(0.25))

    model.add(Conv2D(64,(3,3),activation='relu'))
    model.add(BatchNormalization())
    model.add(Conv2D(64,(3,3),activation='relu'))
    model.add(BatchNormalization())
```

```python
        model.add(MaxPooling2D(pool_size=(2,2)))
        model.add(Dropout(0.25))

        #展开为一维数组全连接层
        model.add(Flatten())
        model.add(Dense(256,activation='relu'))
        model.add(Dropout(0.25))
        model.add(Dense(10))
        return model
```

编写训练函数程序：

```python
def train():
    #载入数据管道
    print("载入数据集>>>")
    train_ds,test_ds=load_dataset()
    print("数据集载入完成!")
    print("载入模型>>>")
    model=build_model()
    model.summary()
    #损失函数为交叉熵损失,优化器为Adam,评估指标为准确率
    loss=tf.keras.losses.SparseCategoricalCrossentropy(from_logits=True)
    optimizer=tf.keras.optimizers.Adam(learning_rate=0.0005)
    model.compile(loss=loss,optimizer=optimizer,metrics=['accuracy'])

    #训练过程
    print("开始训练模型>>>")
    os.makedirs(os.path.dirname(CKPT_PATH))
    steps_per_epoch=len(all_images_path(TRAIN_PATH))//BATCH_SIZE
    validation_steps=len(all_images_path(TEST_PATH))//BATCH_SIZE
    tensorboard_callback=tf.keras.callbacks.TensorBoard(log_dir='./logs',histogram_freq=1)
    checkpoint_callback=tf.keras.callbacks.ModelCheckpoint(CKPT_PATH)
    model.fit(train_ds,
              epochs=50,
              workers=1,
              validation_data=test_ds,
              steps_per_epoch=steps_per_epoch,
              validation_steps=validation_steps,
              callbacks=[tensorboard_callback,checkpoint_callback])
    #保存整个模型
    model.save(MODEL_PATH,save_format="h5")
    print("模型训练完成!")

if __name__=='__main__':
    train()
```

程序清单 task6.py 的完整源代码文件参见 U2-task6.py.pdf。

task6.py

命令行输入命令运行 task6.py，结果如图 2.43～图 2.45 所示。

```
python task6.py
```

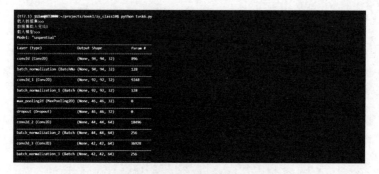

图 2.43　训练模型结果（一）

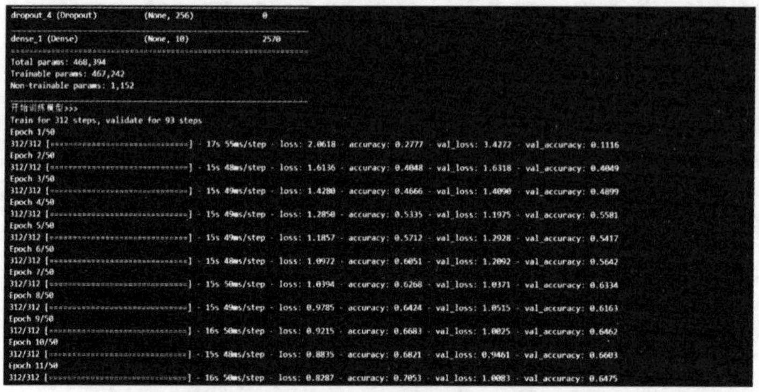

图 2.44　训练模型结果（二）

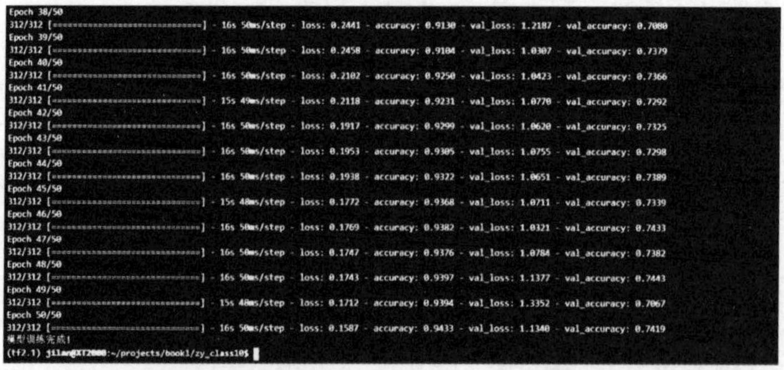

图 2.45　训练模型结果（三）

步骤4：TensorBoard 可视化训练日志。

用 TensorBoard 打开训练日志 logs，命令行输入命令，在浏览器新的标签页输入网址 http://172.16.33.106:8888/，访问 TensorBoard 页面，如图 2.46 所示。

```
tensorboard --host 172.16.33.106 --port 8888 --logdir logs
```

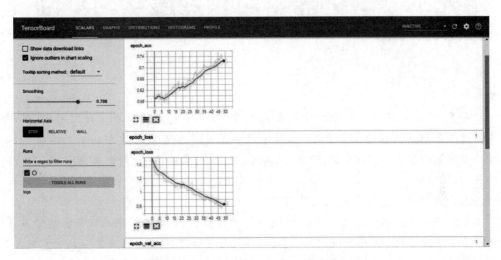

图 2.46　TensorBoard 可视化训练日志

小　结

本单元让读者认识深度学习开发环境，了解相关软硬件组成，学习使用集成开发环境。通过深度学习框架的简介让读者进一步了解深度学习的发展，从 Python 和 Anaconda3 环境简介，让读者掌握 Python 编程基础并能够熟悉 Anaconda3 所配置的集成环境，能够使用 Anaconda3 管理各个独立并存的虚拟环境和环境下的安装包。在 TensorFlow 基础一节，读者需要熟练掌握某一特定版本的 TensorFlow，能够配合使用底层模块和高级模块完成数据处理和构建神经网络模型，在不断编程实践中提升各方面的综合能力。

练　习

练习1　使用 matplotlib 绘制 $y=\sin(x)$ 正弦和 $y=\cos(x)$ 余弦曲线图。

练习2　使用 TensorFlow 计算 $Y=W_2(W_1*X+b_1)+b_2$，其中 $W_1.\text{shape}=(32,10)$，$b_1.\text{shape}=(32,1)$，$W_2.\text{shape}=(128,32)$，$b_1.\text{shape}=(128,1)$，并使用 TensorBoard 可视化数据流图。

练习3　使用 TensorFlow 进行变量实例编程，在构建计算图中区别 tf.Variable() 和 tf.get_variable() 的用法。

练习4　使用 Keras 高级模块构建简单的神经网络模型，完成 zy_dog10 图像数据的分类，输入分辨率为 $256\times256\times3$，输出为 10 种狗的类别。数据集下载地址可以从数据资源平台数据集页面获取。

单元3 机器学习和深度学习基础

近年来,随着大数据和硬件加速设备的发展,深度学习已逐渐成为处理图像、声音和文本等复杂高维度数据的主要方法,机器学习和深度学习等概念焕发出新的生机。国内外诸多互联网科技巨头(如谷歌、脸书、亚马逊、百度、阿里巴巴等)纷纷宣布了将人工智能作为他们的战略重心。在类似AlphaGo、无人驾驶汽车等前沿技术的背后,机器学习和深度学习基础是推动这些技术发展的基石。

本单元将从机器学习的主要任务、经典机器学习算法、深度学习基础以及深度学习算法四个方面帮助读者构建整体性知识框架,要求读者掌握机器学习和深度学习相关基础知识,并能够灵活运用所学算法解决简单的分类和回归问题。本单元的知识导图如图3.1所示。

图 3.1 知识导图

课程安排

课程任务	课程目标	安排课时
机器学习的主要任务	掌握机器学习相关基础概念及机器学习能够解决的主要任务	2
机器学习算法	熟悉并掌握机器学习常用算法类型,并对算法的实现加以了解,能够灵活运用相应算法解决实际问题	3
深度学习基础	掌握深度学习的基本概念,由单层神经网络延伸到多层神经网络,并通过多层感知机引入深度学习模型	3
深度学习算法	对经典深度学习模型加以了解,通过深度学习应用案例对所学内容学以致用	2

3.1 机器学习的主要任务

3.1.1 监督学习

监督学习利用一组已知预期结果的样本调整模型参数,使其达到能准确推断出预期样本的任务。在监督学习中,每个实例都是由一个输入对象和一个期望的输出值组成。监督学习的任务过程就是分析训练数据,建立一个可以产生期望值结果的模型,且该模型可以用于新数据上,如图 3.2 所示。监督学习的应用场景主要有分类问题和回归问题。常见的监督学习算法包括 K 近邻、朴素贝叶斯、线性回归、支持向量机、K 均近聚类等。

(a) 根据已知数据集做训练

(b) 对未知数据集合做分类(预测)

图 3.2 监督学习

3.1.2 无监督学习

无监督学习的输入数据没有被标记,也没有确定的结果,根据这些无标记数据的相关性分析产生结果的过程就是无监督学习。实际应用中,不少情况下无法预先知道样本的标签,也就是说缺乏训练样本对应的类别,因而只能从原先没有样本的标签数据集构建可以产生结果的模型,如图 3.3 所示。常见的应用场景有聚类等。常用的无监督算法主要有 K 均值聚类等。

图 3.3　无监督学习

3.1.3　分类

分类就是将样本数据集划为有限的类别的过程。分类问题是监督学习的主要任务之一,分类输出期望类别结果总是一个有限值,如将[花,鸟,虫,鱼,草],按照[动物,植物]两个类别进行分类,就可以分为[鸟,虫,鱼]和[花,草]。从数据中学习一个分类模型或分类决策函数,这个模型或者决策函数可称为分类器。

分类一般处理的数据集是离散非数值型的数据。机器学习概念下的分类问题包括学习和预测两个过程。在学习过程中,根据已知的训练数据集利用有效的学习方法构建一个分类器;在预测过程中,利用学习的分类器对新的输入数据进行分类,得到分类结果,如图 3.4 所示。

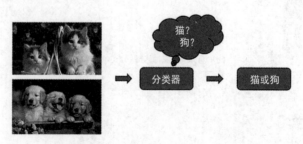

图 3.4　分类

分类任务广泛地应用在多数场合下，分门别类在生活中随处可见，如电影分类、语言分类、动植物分类、图书分类、汽车分类等。

分类评估方法的主要功能是评估分类器算法的好坏，它包括多项指标。了解各种评估方法，在实际应用中选择正确的评估方法是十分重要的，以下列出了常见的评估指标。

1. 正确率（accuracy）

正确率是最常见的评估指标，且很容易计算：对于给定的测试数据集，分类器正确分类的样本数占总测试样本数的百分比。通常来说，正确率越高，分类器越好。

2. 错误率（error rate）

错误率则与正确率相反，描述被分类器错分的比例。

其他评价指标：

- 计算速度（time）：分类器训练和预测需要的时间。
- 鲁棒性（robust）：处理缺失值和异常值的能力。
- 敏感度（sensitivity）：表示所有正例中被分正确的比例，衡量了分类器对正例的识别能力。
- 特异性（specificity）：表示所有负例中被分正确的比例，衡量了分类器对负例的识别能力。

3.1.4 回归

回归是建立预测输入变量和输出变量之间的关系，当输入变量的值发生变化时，输出变量的值随之发生变化。从输入变量到输出变量之间映射的函数模型就是回归器，回归是监督学习的另一个重要任务。

回归一般处理的是连续数值型数据。机器学习概念下回归问题同样包括学习和预测两个过程。在学习过程中，根据已知的训练数据集构建一种映射关系，也就是回归器；在预测过程中，利用学习的回归器对新的输入数据进行计算，得到数据结果，如图3.5所示。

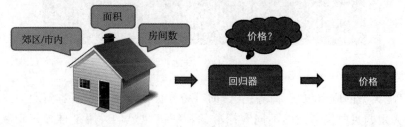

图 3.5　回归

回归问题按照输入变量的个数，分为一元回归和多元回归；按照输入变量和输出变量之间关系的类型（即模型的类型）分为线性回归和非线性回归。许多领域的任务都可以形式化为回归问题，例如，回归可以用于商务领域，进行市场趋势预测、股价预测。

回归学习最常用的评估方法是均方误差、均方根误差以及平均绝对误差。

1. 均方误差（MSE）

均方误差是预测值与真实值之差的平方，反映预值的变化程度，MSE的值越小，说明预测模型描述实验数据具有更好的精确度。

2. 均方根误差(RMSE)

均方根误差是均方误差的算数平方根。

3. 平均绝对误差(MAE)

平均绝对误差是预测值与真实值差的绝对值的平均数,反映预测值误差的实际情况。

3.1.5 聚类

聚类就是对大量未知标注的数据集,按数据的内在相似性将数据集划分为多个类别,使类别内的数据相似度较大,而类别间的数据相似度较小。机器学习中的聚类是无监督的学习问题,它的目标是为了感知样本间的相似度进行类别归纳。简单地说就是把相似的东西分到一组,聚类的时候,并不关心某一类是什么,需要实现的目标只是把相似的东西聚到一起。一个聚类算法通常只需要知道如何计算相似度就可以开始工作了,因此聚类通常并不需要使用训练数据进行学习。以图3.6为例,这里有一堆水果,但事先没有告诉你有哪些水果,也没有一个训练好的判定各种水果的模型,聚类算法要自动将这堆水果进行归类,即相同的水果聚成一堆。

图3.6 聚类

聚类算法典型的应用包括:商务上,帮助市场分析人员从客户基本资料库中发现不同的客户群,并用购买模式来刻画不同客户群的特征;生物学上,用于推导植物和动物的分类,对基因进行分类,获得对种群固有结构的认识;地理信息方面,在地球观测数据库中相似区域的确定、根据房子的地理位置对一个城市中房屋的分组;聚类也能用于对文档进行分类。此外,聚类分析可以作为其他数据挖掘算法的预处理步骤,便于这些算法在生成的簇上进行处理。

聚类算法的主要评估方法有外部有效性评估、内部有效性评估和相关性测试评估。

3.2 机器学习算法

3.2.1 K-近邻算法

1. 算法原理

K-近邻算法的工作原理是:存在一个样本数据集合(又称训练样本集),并且样本集中每个数据都存在标签,即知道样本集中每一数据与所属分类的对应关系。输入没有标签的新数据后,将新数据的每个特征与样本集中数据对应的特征进行比较,然后用算法提取样本集中特征最相似数据(最近邻)的分类标签。一般来说,只选择样本集中前 K 个最相似的数据,这就是 K-近邻算法中 K 的出处。最后,选择 K 个最相似数据中出现次数最多的分类,作为新数据的分类。

(1)优点
①思想简单,理论成熟,既可以用来做分类也可以用来做回归。
②可用于非线性分类。
③训练时间复杂度为 $O(n)$。
④准确度高,对数据没有假设,对奇异值不敏感。

(2)缺点
①计算量较大。
②对于样本分类不均衡的问题,会产生误判。
③需要大量的内存。
④输出的可解释性不强。

2. 应用领域

相似度分析、推荐系统、模式识别、多分类等领域。

单元任务 7　使用 K-近邻识别手写数字

在本例中,使用 tf1.15 虚拟环境,使用 K-近邻算法完成手写数字识别任务。MNIST 手写体字库包含 70 000 张 0~9 手写灰度图,其中 60 000 张标注为训练样本数据集,10 000 张为测试样本数据集。

步骤 1:导入必要的库。

```
import random
import numpy as np
import tensorflow as tf
import matplotlib.pyplot as plt
from PIL import Image
from tensorflow.examples.tutorials.mnist import input_data
from tensorflow.python.framework import ops
ops.reset_default_graph()

sess = tf.Session()
```

步骤 2:数据载入,训练集、测试集拆分。

```
mnist = input_data.read_data_sets("MNIST_data/",one_hot = True)

np.random.seed(13)
train_size = 1000
test_size = 100
rand_train_indices = np.random.choice(len(mnist.train.images),train_size,replace = False)
rand_test_indices = np.random.choice(len(mnist.test.images),test_size,replace = False)
x_vals_train = mnist.train.images[rand_train_indices]
x_vals_test = mnist.test.images[rand_test_indices]
y_vals_train = mnist.train.labels[rand_train_indices]
y_vals_test = mnist.test.labels[rand_test_indices]
```

步骤 3:定义 K 值和批量大小。

```
k = 4
batch_size = 10

x_data_train = tf.placeholder(shape = [None,784],dtype = tf.float32)
x_data_test = tf.placeholder(shape = [None,784],dtype = tf.float32)
y_target_train = tf.placeholder(shape = [None,10],dtype = tf.float32)
y_target_test = tf.placeholder(shape = [None,10],dtype = tf.float32)
```

步骤 4:选择距离度量(这里使用曼哈顿距离)。

```
distance = tf.reduce_sum(tf.abs(tf.subtract(x_data_train,tf.expand_dims(x_data_test,1))),axis = 2)
```

步骤 5:近邻计算。

```
#近邻计算
top_k_xvals,top_k_indices = tf.nn.top_k(tf.negative(distance),k = k)
prediction_indices = tf.gather(y_target_train,top_k_indices)

#类别预测
count_of_predictions = tf.reduce_sum(prediction_indices,axis = 1)
prediction = tf.argmax(count_of_predictions,axis = 1)

#计算循环次数
num_loops = int(np.ceil(len(x_vals_test)/batch_size))
test_output = []
actual_vals = []
for i in range(num_loops):
    min_index = i * batch_size
    max_index = min((i + 1) * batch_size,len(x_vals_train))
    x_batch = x_vals_test[min_index:max_index]
    y_batch = y_vals_test[min_index:max_index]
    predictions = sess.run(prediction,feed_dict = {x_data_train: x_vals_train,x_data_test: x_batch,y_target_train: y_vals_train,y_target_test: y_batch})
```

```
        test_output.extend(predictions)
        actual_vals.extend(np.argmax(y_batch,axis=1))

    accuracy = sum([1./test_size for i in range(test_size) if test_output[i] == 
actual_vals[i]])
    print('测试集准确率:' + str(accuracy))
```

步骤6:可视化。

```
    actuals = np.argmax(y_batch,axis=1)

    Nrows = 2
    Ncols = 3
    for i in range(6):
        plt.subplot(Nrows,Ncols,i+1)
        plt.imshow(np.reshape(x_batch[i],[28,28]),cmap='Greys_r')
        plt.title('Actual:' + str(actuals[i]) + ';' + 'Pred:' + str
(predictions[i]),fontsize=10)
        frame = plt.gca()
        frame.axes.get_xaxis().set_visible(False)
        frame.axes.get_yaxis().set_visible(False)
    plt.savefig('kNN.png')
```

运行结果如图3.7所示。

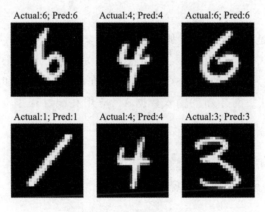

图3.7　手写体分类结果

3.2.2　朴素贝叶斯

1. 算法原理

朴素贝叶斯是基于贝叶斯定理与特征条件独立性假设的一种分类算法,属于生成式模型,对联合分布建模。对于给定的训练数据集 D,首先基于条件独立性假设,学习输入 X 和输出 Y 的联合概率分布 $P(X,Y)$,然后基于此模型,对于给定的输入 x,利用贝叶斯定理求出后验概率 $P(Y|X)$ 最大的输出 y。

(1)优点

①朴素贝叶斯模型发源于古典数学理论,有着坚实的数学基础,以及稳定的分类效率。

②在大数量训练和查询时具有较高的速度。即使使用超大规模的训练集,针对每个项目通常也只会有相对较少的特征数,并且对项目的训练和分类也仅仅是特征概率的数学运算而已。

③对小规模的数据表现很好,能够处理多分类任务,适合增量式训练,实时地对新增样本进行训练。

④朴素贝叶斯对结果解释容易理解。

(2)缺点

①需要计算先验概率。

②分类决策存在错误率。

③对输入数据的表达形式很敏感。

④由于使用了样本属性独立性的假设,所以如果样本属性有关联时其效果不好。

2. 应用领域

欺诈检测、垃圾邮件检测、文章分类等领域。

3.2.3 线性回归

1. 算法原理

如果一个数据集是线性可分的,那么这个数据集就可以用线性函数 $Y = WX + b$ 来描述,其中 X 是输入数据,Y 是输出数据,W 为权重,b 为偏置。线性回归的目的是寻找合适的 W 和 b,用线性函数为原始数据建立线性模型。在计算权重 W 和偏置 b 的过程中,最常用的算法是梯度下降法。对于新的输入数据 X_2,应用在线性模型 $Y = WX + b$ 中得到的 Y_2 值就是预测值。

(1)优点

①实现简单,广泛应用于数值型问题上。

②计算代价不高,易于理解和实现。

(2)缺点

①当特征空间很大时,线性回归的性能不是很好。

②容易欠拟合,不能很好地处理大量多类特征或变量。

③训练数据必须线性可分。

2. 应用领域

在数值型线性数据预测时广泛应用,如房价预测、车辆位置预测等。

单元任务8 使用线性回归预测房价

有一个房价数据集,短期内随着时间呈线性变化,需要对此数据集使用线性回归建立模型,并经过梯度下降算法训练模型的权重和偏置参数,达到能在一定时间内预测房价的目标。

步骤1:构造房价数据。

首先需要创建一个呈线性的房价-时间训练集数据,利用线性函数和随机噪声点创建,并在二维平面可视化数据,其中横坐标为时间,纵坐标为房价,这里构造1 000个样本数据作为训练集。在本任务中使用tf2.3环境进行编程,新建程序linear_regression.py文件。

```
#构建线性数据
num_points = 1000
train_dataset = []
np.random.seed(1)
for i in range (num_points):
    x1 = np.random.uniform(low = 0.0,high = 10)
    y1 = x1* 0.2 + 0.8 + np.random.normal(0.0,0.3)
    train_dataset.append([x1,y1])
train_x = [value[0] for value in train_dataset]
train_y = [value[1] for value in train_dataset]
plt.scatter (train_x,train_y,c = 'blue',s = 8)
plt.xlabel('time/days')
plt.ylabel('house price/10k')
plt.savefig('./house_price.png')
```

生成的随机线性数据如图3.8所示。

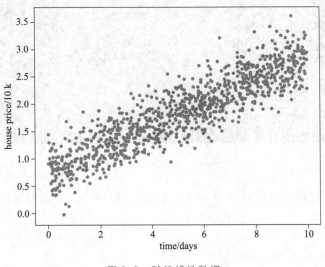

图3.8 随机线性数据

步骤2:数据预处理。

将原有的数据进行乱序处理。

```
#打乱数据
shuffle_index = np.random.permutation(num_points)
x_ds = train_x[shuffle_index]
y_ds = train_y[shuffle_index]
```

步骤3:初始化权重和偏置值。

线性模型的权重和偏置需要指定一个初始值用于后续训练,这里对权重和偏置值进行随机初始化,选用[0,1]间的随机值。

```
initializer = tf.random_uniform_initializer(0.0,1.0)
W = tf.Variable(initializer(shape = [1],dtype = tf.float32),name = 'W')
b = tf.Variable(initializer(shape = [1],dtype = tf.float32),name = 'b')
```

步骤4:使用梯度下降算法更新参数。

在1 000个训练集样本的迭代中,以预测值pred_y和实际值train_y之间的均方差作为损失,使用梯度下降算法更新线性回归模型的权重和偏置,寻找最优化的W和b值,并将训练好的线性回归模型在样本数据点上可视化,参数训练过程如图3.9所示。

```
#采用批量训练
def train_step(epoch,batch,x_batch,y_batch):
    with tf.GradientTape() as tape:
        pred_y = W* x_batch + b
        loss = tf.reduce_mean(tf.math.square(pred_y - y_batch))
    train_variable = [W,b]
    gradient = tape.gradient(loss,train_variable)
    optimizer.apply_gradients(zip(gradient,train_variable))
    print('Epoch:% d \tStep:% 5d \tW:% .3f \tb:% .3f' % (epoch,batch,W.numpy(),b.numpy()))
```

图3.9 参数训练过程

使用批量训练,训练结束后将 W 和 b 在数据点上可视化,结果如图3.10所示。

```
#训练循环
def train():
    for epoch in range(EPOCH):
        for i in range(0,1000,BATCH_SIZE):
            x_batch = x_ds[i: i + BATCH_SIZE]
            y_batch = y_ds[i: i + BATCH_SIZE]
            train_step(epoch,i,x_batch,y_batch)
    print("训练完成!")
    print('W: % .3f \tb:% .3f' % (W.numpy(),b.numpy()))
    plt.scatter(train_x,train_y,c = 'blue',s = 8)
    plt.plot(train_x,W.numpy() * train_x + b.numpy(),c = 'red')
    plt.xlabel('time/days')
    plt.ylabel('house price/10k')
    plt.savefig('predict.png')
```

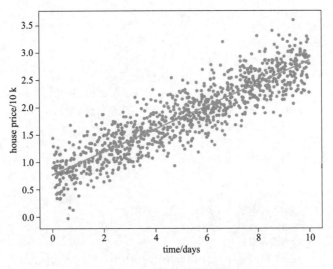

图 3.10 线性拟合结果

步骤 5：完整代码和训练结果。

```
#linear_regression.py
import os
import numpy as np
import tensorflow as tf
import matplotlib.pyplot as plt

EPOCH = 2000
BATCH_SIZE = 100
os.environ['CUDA_VISIBLE_DEVICES'] = '0'
os.environ['TF_CPP_MIN_LOG_LEVEL'] = '3'
os.environ['TF_FORCE_GPU_ALLOW_GROWTH'] = 'true'

num_points = 1000
train_dataset = []
np.random.seed(1)
for i in range(num_points):
    x1 = np.random.uniform(low=0.0, high=10)
    y1 = x1 * 0.2 + 0.8 + np.random.normal(0.0, 0.3)
    train_dataset.append([x1, y1])
train_x = [value[0] for value in train_dataset]
train_y = [value[1] for value in train_dataset]
#plt.scatter(train_x, train_y, c='blue', s=8)
#plt.xlabel('time/days')
#plt.ylabel('house price/10k')
#plt.savefig('./house_price.png')

#打乱数据
shuffle_index = np.random.permutation(num_points)
x_ds = np.array(train_x)
```

```python
y_ds = np.array(train_y)
x_ds = x_ds[shuffle_index]
y_ds = y_ds[shuffle_index]

initializer = tf.random_uniform_initializer(0.0,1.0)
W = tf.Variable(initializer(shape = [1],  dtype = tf.float32),name = 'W')
b = tf.Variable(initializer(shape = [1],dtype = tf.float32),name = 'b')

#梯度下降优化器
optimizer = tf.optimizers.SGD(learning_rate = 0.01)

#采用批量训练
def train_step(epoch,batch,x_batch,y_batch):
    with tf.GradientTape() as tape:
        pred_y = W* x_batch + b
        loss = tf.reduce_mean(tf.math.square(pred_y - y_batch))
    train_variable = [W,b]
    gradient = tape.gradient(loss,train_variable)
    optimizer.apply_gradients(zip(gradient,train_variable))
    print('Epoch:% d\tStep:% 5d\tW:% .3f\tb:% .3f' %  (epoch,batch,W.numpy(),b.numpy()))

def train():
    for epoch in range(EPOCH):
        for i in range(0,1000,BATCH_SIZE):
            x_batch = x_ds[i: i + BATCH_SIZE]
            y_batch = y_ds[i: i + BATCH_SIZE]
            train_step(epoch,i,x_batch,y_batch)
    print("训练完成!")
    print('W:% .3f\tb:% .3f' %  (W.numpy(),b.numpy()))
    plt.scatter(train_x,train_y,c = 'blue',s = 8)
    plt.plot(train_x,W.numpy()* train_x + b.numpy(),c = 'red')
    plt.xlabel('time/days')
    plt.ylabel('house price/10k')
    plt.savefig('predict.png')

if __name__ == '__main__':
    train()
```

3.2.4 支持向量机

1. 算法原理

支持向量的意思就是数据集中的某些位置特殊的点。例如,存在直线 $x + y - 2 = 0$,直线上面区域($x + y - 2 > 0$)的全是 A 类,直线下面区域的($x + y - 2 < 0$)的全是 B 类。确定这条直线时,主要观察聚集在一起的两类数据中各自最边缘位置的点,也就是最靠近划分直线的那几个点,而其他点对直线最终位置的确定起不了作用,所以称这些点为"支持向量"。运用支持向量分类的模型称为支持向量机(Support Vector Machines, SVM)。

（1）优点

①SVM 是一种有坚实理论基础的新颖的适用小样本学习的方法。
②计算的复杂性取决于支持向量的数目,而不是样本空间的维数。
③少数支持向量决定了最终结果,对异常值不敏感,有优秀的泛化能力。

（2）缺点

①大规模训练难以实施。
②解决多分类问题困难。
③对参数和核函数选择敏感。

2. 应用领域

文本分类、图像分类、人脸分类。

单元任务 9　使用支持向量机实现鸢尾花分类

本任务将使用支持向量机实现鸢尾花数据集的分类,鸢尾花数据集是机器学习和统计分析最经典的数据集之一。鸢尾花(Iris)有三个亚属,分别是山鸢尾(Iris-setosa)、变色鸢尾(Iris-versicolor)和维吉尼亚鸢尾(Iris-virginica),可以通过花萼和花瓣的长度及宽度作为特征,实现鸢尾花的分类。总共有 150 个数据集,每类有 50 个样本。本例将使用花萼宽度和花萼长度的特征创建一个线性二值分类器来预测是否为山鸢尾花。

步骤 1:导入必要的库。

本任务使用 tf1.15 环境编写代码。

```
import matplotlib.pyplot as plt
import numpy as np
import tensorflow as tf
from sklearn import datasets
from tensorflow.python.framework import ops

os.environ['CUDA_VISIBLE_DEVICES'] = '0'
os.environ['TF_CPP_MIN_LOG_LEVEL'] = '3'
os.environ['TF_FORCE_GPU_ALLOW_GROWTH'] = 'true'
ops.reset_default_graph()
sess = tf.Session()
```

步骤 2:载入数据。

```
#载入数据
iris = datasets.load_iris()
x_vals = np.array([[x[0],x[3]] for x in iris.data])
y_vals = np.array([1 if y==0 else -1 for y in iris.target])
class1_x = [x[0] for i,x in enumerate(x_vals) if y_vals[i]==1]
class1_y = [x[1] for i,x in enumerate(x_vals) if y_vals[i]==1]
class2_x = [x[0] for i,x in enumerate(x_vals) if y_vals[i]==-1]
class2_y = [x[1] for i,x in enumerate(x_vals) if y_vals[i]==-1]
```

步骤 3:设置模型参数。

```
#定义批量大小
batch_size =150
```

```
x_data = tf.placeholder(shape = [None,2],dtype = tf.float32)
y_target = tf.placeholder(shape = [None,1],dtype = tf.float32)
prediction_grid = tf.placeholder(shape = [None,2],dtype = tf.float32)

b = tf.Variable(tf.random.normal(shape = [1,batch_size]))
```

步骤4:选择核函数。

对于二维线性可分的数据,可以轻易地计算出数据的分离面,而对于二维线性不可分的数据,通常需要将数据映射到高维空间再计算分离超平面,这种映射关系需要一个函数来确定,这就是核函数。这里选择高斯核函数。

```
#高斯核函数
gamma = tf.constant(-50.0)
sq_vec = tf.multiply(2.,tf.matmul(x_data,tf.transpose(x_data)))
my_kernel = tf.exp(tf.multiply(gamma,tf.abs(sq_vec)))
```

步骤5:定义损失函数。

```
#损失函数定义
first_term = tf.reduce_sum(b)
b_vec_cross = tf.matmul(tf.transpose(b),b)
y_target_cross = tf.matmul(y_target,tf.transpose(y_target))
second_term = tf.reduce_sum(tf.multiply(my_kernel,tf.multiply(b_vec_cross,y_target_cross)))
loss = tf.negative(tf.subtract(first_term,second_term))
```

步骤6:定义预测函数和准确度函数,用来评估训练集和测试集训练的准确度。

```
#预测函数和准确度函数
rA = tf.reshape(tf.reduce_sum(tf.square(x_data),1),[-1,1])
rB = tf.reshape(tf.reduce_sum(tf.square(prediction_grid),1),[-1,1])
pred_sq_dist = tf.add(tf.subtract(rA,tf.multiply(2.,tf.matmul(x_data,tf.transpose(prediction_grid)))),tf.transpose(rB))
pred_kernel = tf.exp(tf.multiply(gamma,tf.abs(pred_sq_dist)))

prediction_output = tf.matmul(tf.multiply(tf.transpose(y_target),b),pred_kernel)
prediction = tf.sign(prediction_output - tf.reduce_mean(prediction_output))
accuracy = tf.reduce_mean(tf.cast(tf.equal(tf.squeeze(prediction),tf.squeeze(y_target)),tf.float32))
```

步骤7:定义优化函数。

```
#定义优化函数
my_opt = tf.train.GradientDescentOptimizer(0.01)
train_step = my_opt.minimize(loss)

#变量初始化
init = tf.global_variables_initializer()
sess.run(init)
```

步骤8：模型训练。

```
#训练过程
loss_vec = []
batch_accuracy = []
for i in range(300):
    rand_index = np.random.choice(len(x_vals), size = batch_size)
    rand_x = x_vals[rand_index]
    rand_y = np.transpose([y_vals[rand_index]])
    sess.run(train_step, feed_dict = {x_data: rand_x, y_target: rand_y})

    temp_loss = sess.run(loss, feed_dict = {x_data: rand_x, y_target: rand_y})
    loss_vec.append(temp_loss)

    acc_temp = sess.run(accuracy, feed_dict = {x_data: rand_x,
                                               y_target: rand_y,
                                               prediction_grid: rand_x})
    batch_accuracy.append(acc_temp)

    if (i + 1) % 75 == 0:
        print('Step:' + str(i + 1) + '\t' + 'Loss: ' + str(temp_loss))
```

步骤9：可视化展示。

```
x_min, x_max = x_vals[:, 0].min() - 1, x_vals[:, 0].max() + 1
y_min, y_max = x_vals[:, 1].min() - 1, x_vals[:, 1].max() + 1
xx, yy = np.meshgrid(np.arange(x_min, x_max, 0.02),
                     np.arange(y_min, y_max, 0.02))
grid_points = np.c_[xx.ravel(), yy.ravel()]
[grid_predictions] = sess.run(prediction, feed_dict = {x_data: x_vals,
                                                       y_target: np.transpose([y_vals]),
                                                       prediction_grid: grid_points})
grid_predictions = grid_predictions.reshape(xx.shape)

plt.contourf(xx, yy, grid_predictions, cmap = plt.cm.Paired, alpha = 0.8)
plt.plot(class1_x, class1_y, 'ro', label = 'Iris setosa')
plt.plot(class2_x, class2_y, 'bo', label = 'Non setosa')
plt.title('Gaussian SVM Results on Iris Data')
plt.xlabel('Petal Length')
plt.ylabel('Sepal Width')
plt.legend(loc = 'lower right')
plt.ylim([-0.5, 3.0])
plt.xlim([3.5, 8.5])
plt.savefig('iris_svm_data.png')
```

结果如图3.11所示。

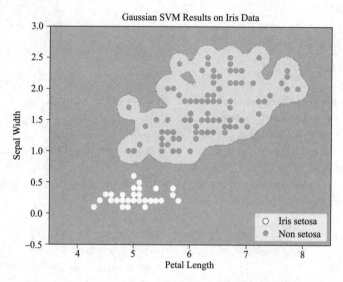

图 3.11 支持向量机分类结果

```
#绘制准确率图像,如图 3.12 所示
plt.plot(batch_accuracy,label = 'Accuracy')
plt.title('Batch Accuracy')
plt.xlabel('Generation')
plt.ylabel('Accuracy')
plt.legend(loc = 'lower right')
plt.savefig('iris_svm_acc.png')

#绘制损失图像,如图 3.13 所示
plt.plot(loss_vec)
plt.title('Loss per Generation')
plt.xlabel('Generation')
plt.ylabel('Loss')
plt.savefig('iris_svm_loss.png')
```

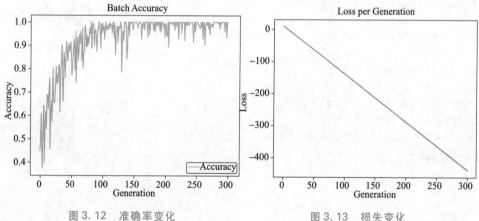

图 3.12 准确率变化　　　　　　　　图 3.13 损失变化

3.2.5 K-均值聚类

1. 算法原理

聚类是一种无监督的学习算法,它将相似的对象归到一个簇中,将不相似对象归到不同簇中。相似这一概念取决于所选择的相似度计算方法。K-均值聚类是发现给定数据集的 K 个簇的聚类算法,之所以称为 K-均值是因为它可以发现 K 个不同的簇,且每个簇的中心采用簇中所含值的均值计算而成。簇个数 K 是用户指定的,每一个簇通过其质心,即簇中所有点的中心来描述。聚类与分类算法的最大区别在于,分类的目标类别已知,而聚类的目标类别是未知的。

① 优点:容易实现。
② 缺点:可能收敛到局部最小值,在大规模数据集上收敛较慢。

2. 应用领域

主要用来聚类,但是类别是未知的,如图像分割、商业选址等。

3.3 深度学习基础

3.3.1 神经网络

人工神经网络模型是深度学习的核心,因其在广泛的应用场景下具有通用性、强大性和可扩展性,使得它们能够很好地解决大型和高度复杂的机器学习任务,例如图像分类任务、语音识别任务、机器翻译任务、通过追踪用户行为的推荐系统,以及通过深度强化学习实现智能游戏专家系统。下面介绍深度学习的基础知识,让读者对人工神经网络建立一个基本的认知。

1. 生物神经元

人工神经元(ANN)受启发于生物神经元,生物神经元是一种异常细胞,大量存在于动物大脑皮层中,如图 3.14 所示,由包含细胞核和大多数细胞复杂成分的细胞体组成,外围的分支扩展称为树突,长的延伸称为轴突。轴突的末端又分裂成许多分支,在这些分支的顶端是微小的结构,称为突触末端,它们连接到其他神经元的树突和其他细胞体。生物神经元接收短的电脉冲,称为来自其他神经元的信号,通过这些突触,当神经元在几毫秒内接收到来自其他神经元的足够数量的信号时,它就发射出自己的信号。

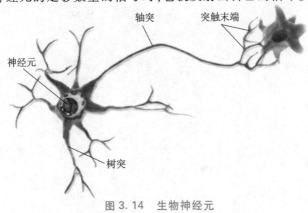

图 3.14 生物神经元

个体的生物神经元似乎以一种相当简单的方式运行,但是它们组织在一个巨大的数十亿神经元的网络中,每个神经元通常连接到数千个其他神经元。高度复杂的计算可以由相当简单的神经元的巨大网络完成。

2. 感知器

感知器是最简单的人工神经网络结构之一,如图3.15所示。

输入:$x_1, x_2, x_3, \ldots, x_n$

加权和:$z = W_1 \times x_1 + W_2 \times x_2 + W_3 \times x_3 + \cdots + W_n \times x_n$

激活:阶跃函数 $\text{step}(z)$

输出:$y = \text{step}(z)$

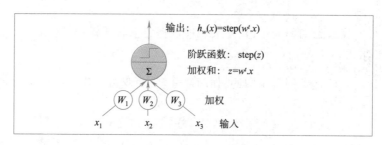

图 3.15 感知器

单层感知器由一层线性阈值单元组成,每个线性阈值单元都连接着若干输入,这些输入通常经过权重和偏置与神经元相连,再被阶跃函数激活,然后输出结果。

堆叠多个层的感知器就形成多层感知机(MLP)。感知机接收一个训练实例,并进行预测,对于每一个产生错误预测的输出神经元,它会加强输入的连接权重,这有助于正确预测。每个输出神经元的决策边界是线性的,因此单层感知器不能学习复杂的模式,但是堆叠多个层的多层感知机可以用来实现复杂模式的预测任务。

3. 深度神经网络与反向传播

多层感知机通常由一个输入层、一个或多个隐藏层、一个输出层组成。除了输出层之外的每一层包括偏置神经元,并且全连接到下一层。当人工神经网络有两个或多个隐含层时,称为深度神经网络(DNN),如图3.16所示。

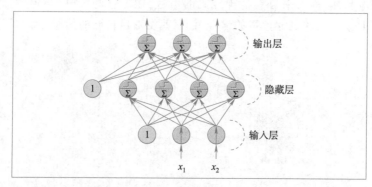

图 3.16 深度神经网络

对于训练实例,将其送到网络并计算每个连续层中每个神经元的输出,这是一次向前传递。然后,它测量网络的输出误差,并且计算最后隐藏层中的每个神经元对每个输出神经元的误差梯度。然后,继续测量这些误差贡献有多少来自先前隐藏层中的每个神经元等,直到算法到达输入层。简单来说,对于每个训练实例,反向传播算法首先进行前向计算,统计误差,然后反向遍历每个层统计每个连接的误差梯度,最后调整网络权值以减少误差。

单元任务10 汽车油耗预测

本任务将利用多层感知机网络模型完成汽车效能指标 MPG 的预测问题,MPG 指标表示汽车燃烧每加仑燃油可行驶的英里数。

步骤1:准备数据集。

采用 Auto MPG 数据集,它记录了各种汽车效能指标与气缸数、质量、马力等因子的真实数据,除了产地字段表示类别外,其他字段都是数值类型。对于产地段,1 表示美国,2 表示欧洲,3 表示日本。本任务采用 tf2.1 的虚拟环境执行程序,tf2.1 环境中需要安装 pandas。

```
#从数据资源服务器下载数据集
wget http://172.16.33.72/dataset/machine_learning_datasets/auto-mpg.data
```

Auto MPG 数据集共记录了 398 项数据,数据集字段含义见表 3.1,从数据资源服务器下载并读取数据集到 DataFrame 对象中。

表 3.1 Auto MPG 数据集表头字段描述

MPG	Cylinders	Displacement	Horsepower	Weight	Acceleration	Model Year	Origin
汽车燃烧每加仑燃油可行驶的英里数	气缸数量	排量	马力	质量	加速度	型号年份	产地

利用 pandas 的 DataFrame 对象加载部分数据集,结果如图 3.17 所示。

```
def load_dataset():
    #表头名称分别为:效能,气缸数,排量,马力,质量,加速度,型号年份,产地
    column_names = ['MPG','Cylinders','Displacement','Horsepower','Weight',
                    'Acceleration','Model Year','Origin']
    raw_dataset =pd.read_csv('./auto-mpg.data',names = column_names,na_values ="?",
comment = '\t',sep =" ",skipinitialspace =True)
    dataset = raw_dataset.copy()
    return dataset

dataset = load_dataset()
print(dataset.head())
```

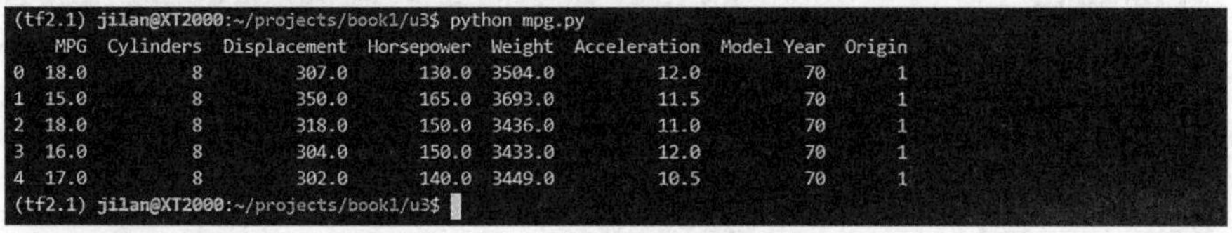

```
(tf2.1) jilan@XT2000:~/projects/book1/u3$ python mpg.py
    MPG  Cylinders  Displacement  Horsepower  Weight  Acceleration  Model Year  Origin
0  18.0          8         307.0       130.0  3504.0          12.0          70       1
1  15.0          8         350.0       165.0  3693.0          11.5          70       1
2  18.0          8         318.0       150.0  3436.0          11.0          70       1
3  16.0          8         304.0       150.0  3433.0          12.0          70       1
4  17.0          8         302.0       140.0  3449.0          10.5          70       1
(tf2.1) jilan@XT2000:~/projects/book1/u3$
```

图 3.17 Auto MPG 部分数据集

步骤 2:数据预处理。

①原始表格中的数据可能含有空字段(缺失值)的数据项,需要清除这些记录项。

②由于 Origin 字段为类别类型数据,将其移除,并转换为 3 个新字段: USA、Europe 和 Japan,分别代表是否来自此产地。

③按照 8:2 的比例切分数据集为训练集和测试集。

④统计训练集各个字段数值的均值和标准差,并完成数据的标准化,数据预处理结果如图 3.18 所示。

```
def preprocess_dataset(dataset):
    dataset = dataset.copy()
    #统计空白数据,并清除
    dataset = dataset.dropna()

    #处理类别型数据,其中 origin 列代表了类别 1、2、3,分别代表产地:美国、欧洲、日本
    #其弹出这一列
    origin = dataset.pop('Origin')
    #根据 origin 列写入新列
    dataset['USA'] = (origin = =1)* 1.0
    dataset['Europe'] = (origin = = 2)* 1.0
    dataset['Japan'] = (origin = =3)* 1.0

    #切分为训练集和测试集
    train_dataset = dataset.sample(frac = 0.8,random_state = 0)
    test_dataset = dataset.drop(train_dataset.index)

    return train_dataset,test_dataset

#查看训练集的输入 X 的统计数据
train_stats = train_dataset.describe()
train_stats = train_stats.traanspose()
print(train_stats)
def norm(x,train_stats):
    return (x-train_stats['mean'])/train_stats['std']

#移动 MPG 油耗效能这一列为真实标签 Y
train_labels = train_dataset.pop('MPG')
test_labels = test_dataset.pop('MPG')

#进行标准化
normed_train_data = norm(train_dataset,train_stats)
normed_test_data = norm(test_dataset,train_stats)
```

```
(tf2.1) jilan@XT2000:~/projects/book1/u3$ python mpg.py
              count         mean         std     min      25%     50%      75%     max
MPG           314.0    23.310510    7.728652    10.0    17.00    22.0    28.95    46.6
Cylinders     314.0     5.477707    1.699788     3.0     4.00     4.0     8.00     8.0
Displacement  314.0   195.318471  104.331589    68.0   105.50   151.0   265.75   455.0
Horsepower    314.0   104.869427   38.096214    46.0    76.25    94.5   128.00   225.0
Weight        314.0  2990.251592  843.898596  1649.0  2256.50  2822.5  3608.00  5140.0
Acceleration  314.0    15.559236    2.789230     8.0    13.80    15.5    17.20    24.8
Model Year    314.0    75.898089    3.675642    70.0    73.00    76.0    79.00    82.0
USA           314.0     0.624204    0.485101     0.0     0.00     1.0     1.00     1.0
Europe        314.0     0.178344    0.383413     0.0     0.00     0.0     0.00     1.0
Japan         314.0     0.197452    0.398712     0.0     0.00     0.0     0.00     1.0
(tf2.1) jilan@XT2000:~/projects/book1/u3$
```

图 3.18 数据预处理

步骤3:构建感知机模型。

构建一个包含三个全连接层的感知机网络。

```python
class Network(Model):
    #回归网络
    def __init__(self):
        super(Network,self).__init__()
        #创建3个全连接层
        self.fc1 = layers.Dense(64,activation = 'relu')
        self.fc2 = layers.Dense(128,activation = 'relu')
        self.fc3 = layers.Dense(1)

    def call(self,inputs):
        #依次通过3个全连接层
        x = self.fc1(inputs)
        x = self.fc2(x)
        x = self.fc3(x)
        return x

def build_model():
    #创建网络
    model = Network()
    model.build(input_shape = (4,9))
    model.summary()
    return model
```

步骤4:训练模型。

编写训练的代码程序并运行,结果如图3.19和图3.20所示。

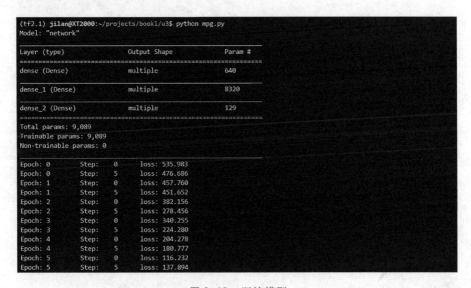

图3.19 训练模型

```
Epoch: 290      Step:    0      loss: 76.505
Epoch: 290      Step:    5      loss: 44.198
Epoch: 291      Step:    0      loss: 50.903
Epoch: 291      Step:    5      loss: 58.954
Epoch: 292      Step:    0      loss: 76.434
Epoch: 292      Step:    5      loss: 50.691
Epoch: 293      Step:    0      loss: 82.806
Epoch: 293      Step:    5      loss: 51.308
Epoch: 294      Step:    0      loss: 68.024
Epoch: 294      Step:    5      loss: 47.943
Epoch: 295      Step:    0      loss: 53.130
Epoch: 295      Step:    5      loss: 53.196
Epoch: 296      Step:    0      loss: 64.940
Epoch: 296      Step:    5      loss: 68.552
Epoch: 297      Step:    0      loss: 43.651
Epoch: 297      Step:    5      loss: 58.840
Epoch: 298      Step:    0      loss: 76.217
Epoch: 298      Step:    5      loss: 56.634
Epoch: 299      Step:    0      loss: 68.481
Epoch: 299      Step:    5      loss: 55.440
训练完成!
(tf2.1) jilan@XT2000:~/projects/book1/u3$
```

图 3.20 训练模型

```
def train(model,train_ds,optimizer,normed_test_data,test_labels):
    train_mae_losses=[]
    test_mae_losses=[]
    for epoch in range(EPOCH):
        for step,(x,y) in enumerate(train_ds):
            with tf.GradientTape() as tape:
                out=model(x)
                loss=tf.reduce_mean(losses.MSE(y,out))
                mae_loss=tf.reduce_mean(losses.MAE(y,out))

            if step % 5==0:
                print('Epoch:% d\tStep:% 5d\tloss: % .3f'% (epoch,step,loss.numpy()))

            grads=tape.gradient(loss,model.trainable_variables)
            optimizer.apply_gradients(zip(grads,model.trainable_variables))

        train_mae_losses.append(float(mae_loss))
        out=model(tf.constant(normed_test_data.values))
        test_mae_losses.append(tf.reduce_mean(losses.MAE(test_labels,out)))

    return train_mae_losses,test_mae_losses
```

步骤5:可视化展示。

将训练集和测试集的损失变化可视化,如图3.21所示。

```
def plot(train_mae_losses,test_mae_losses):
    plt.figure()
    plt.xlabel('Epoch')
    plt.ylabel('MAE')
    plt.plot(train_mae_losses,label='Train')
    plt.plot(test_mae_losses,label='Test')
    plt.legend()
    plt.legend()
    plt.savefig('MPG.png')
```

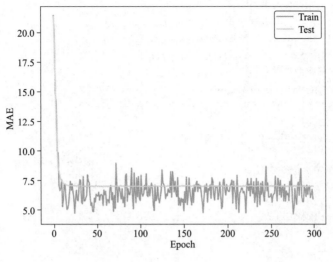

图 3.21 损失变化

步骤 6：完整程序代码。

```
import os
import pandas as pd
import tensorflow as tf
import matplotlib.pyplot as plt
from tensorflow.keras import Model
from tensorflow.keras import layers,losses

EPOCH = 300
BATCH_SIZE = 32
os.environ['CUDA_VISIBLE_DEVICES'] = '0'
os.environ['TF_CPP_MIN_LOG_LEVEL'] = '3'
os.environ['TF_FORCE_GPU_ALLOW_GROWTH'] = 'true'
def load_dataset():
    #表头名称分别为：效能,气缸数,排量,马力,重量,加速度,型号年份,产地
    column_names = ['MPG','Cylinders','Displacement','Horsepower','Weight',
                    'Acceleration','Model Year','Origin']
    raw_dataset = pd.read_csv('./auto-mpg.data', names = column_names,
na_values = "?",comment = '\t',sep = " ",skipinitialspace = True)
    dataset = raw_dataset.copy()
    return dataset


def preprocess_dataset(dataset):
    dataset = dataset.copy()
    #清除空白的无用数据
    dataset = dataset.dropna()
    #处理类别型数据,其中 origin 列代表了类别 1、2、3,分别代表产地:美国、欧洲、日本
    origin = dataset.pop('Origin')
```

```python
    #根据origin列写入新列
    dataset['USA'] = (origin = =1) * 1.0
    dataset['Europe'] = (origin = =2) * 1.0
    dataset['Japan'] = (origin = =3) * 1.0
    #切分为训练集和测试集
    train_dataset = dataset.sample(frac = 0.8, random_state = 0)
    test_dataset = dataset.drop(train_dataset.index)

    return train_dataset, test_dataset

def norm(x, train_stats):
    return (x-train_stats['mean'])/train_stats['std']

class Network(Model):
    #回归网络
    def __init__(self):
        super(Network, self).__init__()

        #创建3个全连接层
        self.fc1 = layers.Dense(64, activation = 'relu')
        self.fc2 = layers.Dense(128, activation = 'relu')
        self.fc3 = layers.Dense(1)

    def call(self, inputs):
        #依次通过3个全连接层
        x = self.fc1(inputs)
        x = self.fc2(x)
        x = self.fc3(x)
        return x

def build_model():
    #创建网络
    model = Network()
    model.build(input_shape = (4,9))
    model.summary()
    return model

def train(model, train_ds, optimizer, normed_test_data, test_labels):
    train_mae_losses = []
    test_mae_losses = []
    for epoch in range(EPOCH):
        for step, (x, y) in enumerate(train_ds):
            with tf.GradientTape() as tape:
                out = model(x)
                loss = tf.reduce_mean(losses.MSE(y, out))
                mae_loss = tf.reduce_mean(losses.MAE(y, out))

            if step % 5 = =0:
                print('Epoch:% d \tStep:% 5d \tloss:% .3f ' %  (epoch, step,
                loss.numpy()))
```

```python
            grads = tape.gradient(loss, model.trainable_variables)
            optimizer.apply_gradients(zip(grads, model.trainable_variables))
        train_mae_losses.append(float(mae_loss))
    out = model(tf.constant(normed_test_data.values))
    test_mae_losses.append(tf.reduce_mean(losses.MAE(test_labels, out)))

    return train_mae_losses, test_mae_losses


def plot(train_mae_losses, test_mae_losses):
    plt.figure()
    plt.xlabel('Epoch')
    plt.ylabel('MAE')
    plt.plot(train_mae_losses, label = 'Train')
    plt.plot(test_mae_losses, label = 'Test')
    plt.legend()
    #plt.ylim([0,10])
    plt.legend()
    plt.savefig('MPG.png')


def main():
    dataset = load_dataset()
    #print(dataset.head())
    train_dataset, test_dataset = preprocess_dataset(dataset)
    train_stats = train_dataset.describe()
    train_stats.pop("MPG")
    train_stats = train_stats.transpose()
    #print(train_stats)
    train_labels = train_dataset.pop('MPG')
    test_labels = test_dataset.pop('MPG')
    #进行标准化
    normed_train_data = norm(train_dataset, train_stats)
    normed_test_data = norm(test_dataset, train_stats)

    train_ds = tf.data.Dataset.from_tensor_slices((normed_train_data.values, train_labels.values))
    train_ds = train_ds.shuffle(300).batch(BATCH_SIZE)
    model = build_model()
    optimizer = tf.keras.optimizers.RMSprop(0.001)
    train_mae_losses, test_mae_losses = train(model, train_ds, optimizer, normed_test_data, test_labels)
    print("训练完成!")
    plot(train_mae_losses, test_mae_losses)


if __name__ == '__main__':
    main()
```

3.3.2 梯度下降法和批处理

1. 梯度下降法

梯度下降法是机器学习中经典的优化算法之一,用于寻求一个曲线的最小值。"梯度"指一条曲线的坡度或倾斜率,"下降"指下降递减的过程。梯度下降法是迭代的,也就是说需要多次计算结果,最终求得最优解。梯度下降的迭代质量有助于使输出结果尽可能拟合训练数据。

在训练模型时,如果训练数据过多,无法一次性将所有数据送入计算,就会遇到批处理问题。为了克服数据量多的问题,选择将数据分成几个部分[即不同批次(batch)]进行训练,从而使得每个批次的数据量都是可以负载的。将这些批次的数据逐一送入计算训练,更新神经网络的权值,使得网络收敛。

2. 批处理

如果数据集比较小,可采用全数据集的形式,其优点是:由全数据集确定的方向能够更好地代表样本总体,从而更准确地朝向极值所在的方向。对于更大的数据集,如果采用全数据集的形式,其缺点是:随着数据集的海量增长和内存限制,一次性载入所有数据变得越来越不可行,并且由于不同权重的梯度值差别巨大,因此选取一个全局的学习率很困难。

假如每次只训练一个样本,线性神经元在均方误差代价函数的错误面是一个抛物面,横截面是椭圆。对于多层神经元、非线性网络,在局部依然近似是抛物面。此时,每次修正方向以各自样本的梯度方向修正,各自为政,难以达到收敛。

既然 Batch_Size 为全数据集或者 Batch_Size = 1 都有各自的缺点,可不可以选择一个适中的 Batch_Size 值呢? 此时,可采用批梯度下降法(Mini-batches Learning)。因为如果数据集足够充分,那么用一部分数据训练算出来的梯度与用全部数据训练出来的梯度几乎是一样的。

3.3.3 损失函数

损失函数(Loss Function)又称误差函数,用来衡量算法的运行情况,估量模型的预测值与真实值的不一致程度,是一个非负实值函数,通常使用 $L(Y,f(x))$ 表示。损失函数越小,模型的鲁棒性就越好。损失函数是经验风险函数的核心部分,也是结构风险函数的重要组成部分。

1. 常见损失函数

深度学习通过对算法中的目标函数进行不断求解优化,得到最终想要的结果。分类和回归问题中,通常使用损失函数或代价函数作为目标函数。

损失函数用来评价预测值和真实值不一样的程度。通常损失函数越好,模型的性能也越好。

损失函数可分为经验风险损失函数和结构风险损失函数。经验风险损失函数指预测结果和实际结果的差别,结构风险损失函数是在经验风险损失函数上加上正则项。

下面介绍常用的损失函数:

(1) 0—1 损失函数

如果预测值和目标值相等,值为 0;如果不相等,值为 1。

$$L(Y,f(x)) = \begin{cases} 1, & Y \neq f(x) \\ 0, & Y = f(x) \end{cases}$$

一般情况下,在实际使用中,相等的条件过于严格,可适当放宽条件:

$$L(Y,f(x)) = \begin{cases} 1, & |Y-f(x)| \geq T \\ 0, & |Y-f(x)| < T \end{cases}$$

(2)绝对值损失函数

和 0—1 损失函数相似,绝对值损失函数表示为:

$$L(Y,f(x)) = |Y-f(x)|$$

(3)平方损失函数

$$L(Y,f(x)) = \sum_N (Y-f(x))^2$$

这点可从最小二乘法和欧几里得距离的角度理解。最小二乘法的原理是,最优拟合曲线应该使所有点到回归直线的距离之和最小。

(4)对数损失函数

$$L(Y,P(Y|X)) = -\log P(Y|X)$$

常见的逻辑回归使用的就是对数损失函数,有很多人认为逻辑回归的损失函数是平方损失,其实不然。逻辑回归假设样本服从伯努利分布(0—1 分布),求得满足该分布的似然函数,然后取对数求极值等。逻辑回归推导出的经验风险函数是最小化的负的似然函数,从损失函数的角度看,就是对数损失函数。

(5)指数损失函数

指数损失函数的标准形式为:

$$L(Y,f(x)) = \exp(-Yf(x))$$

2. 过拟合

在深度学习中,将模型在训练集上的误差称为训练误差,又称经验误差,在新的数据集上的误差称为泛化误差,泛化误差也可以说是模型在总体样本上的误差。对于一个好的模型应该是经验误差约等于泛化误差,也就是经验误差要收敛于泛化误差,根据霍夫丁不等式可知经验误差在一定条件下是可以收敛于泛化误差的。

模型在新的数据集上的表达能力称为泛化能力。当深度模型对训练集学习得太好时,表现为经验误差很小,但泛化误差会很大,这种情况称为过拟合(Overfitting);而当模型在数据集上学习得不够好时,此时经验误差较大,这种情况称为欠拟合(Underfitting)。具体表现如图 3.22 所示。

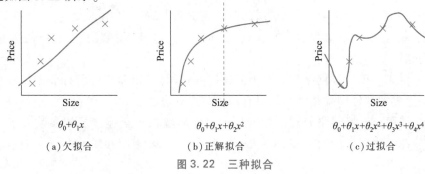

图 3.22 三种拟合

3. 正则化

为了避免过拟合,一种常用的方法是正则化(regularization),简单来说就是修改原来的损失函数,加入模型复杂度衡量的指标,此时,进行优化就是 $J(\theta)+\lambda R(w)$,通常使用的 $R(w)$ 有两种,一种是 L1 正则化,计算方式如下:

$$R(w)=\|w\|_1=\sum|w|$$

另一种是 L2 正则化,计算方式如下:

$$R(w)=\|w\|_2^2=\sum|w|^2$$

两种方式都是通过限制权重的大小,使得模型不能任意地拟合训练数据中的随机噪音,其中 L1 正则化方法会使得参数变得稀疏(也就是会有更多的参数变为 0)并且不能求导,大部分使用的都是 L2 正则化方式。

3.4 深度学习算法

3.4.1 卷积神经网络

1. 卷积神经网络简介

在前面的章节中所介绍的神经网络每两层之间的所有节点都是有边相连的,所以称这种网络结构为全连接网络结构。下面学习一种新的网络结构——卷积神经网络(Convolutional Neural Network,CNN)。

(1)全连接神经网络与 CNN 的异同

①相同点:全连接神经网络和卷积神经网络如图 3.23 所示。

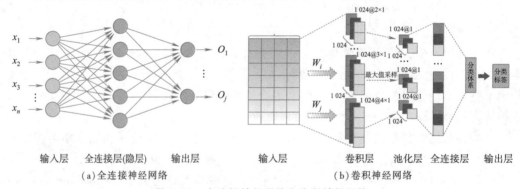

(a)全连接神经网络　　　　　　　　　(b)卷积神经网络

图 3.23　全连接神经网络和卷积神经网络

- 结构相似,二者都是通过一层层的节点组织起来的。
- 输入/输出及训练流程也基本一致,以图像分类为例,输入层都是一维的像素点。

②不同点:

- 唯一区别在于神经网络相邻两层的连接方式,前者相邻两层所有节点都是有边连接,而卷积神经网络相邻两层只有部分节点有边连接。

也因此,使用全连接网络处理图像问题的最大问题在于全连接层的参数太多。对于 MNIST 数据,每一张图片的大小是 $28\times28\times1$,其中 28×28 为图片的大小,×1 表示图像是黑白的,只有一个色彩通道。假设第一层隐藏层的节点数为 500 个,那么一个全连接神经网络将有 28×28×500+500=392 500 个参数。当图片更大时,例如在 CIFAR-10 数据集中,图片的大小为 32×32×3,其中 32×32 表示图片的大小,×3 表示图片是通过红绿蓝三个色彩通道(channel)表示的。这样输入层就有 3 072 个节点,如果第一层全连接层仍然是 500 个节点,那么这一层全连接神经网络将有 3 072×500+500,约 150 万个参数。参数增多除了导致计算速度减慢,还很容易导致过拟合问题。所以需要一个更合理的神经网络结构来有效地减少神经网络中参数的个数,卷积神经网络就可以达到这个目的。

(2)CNN 基本结构

如图 3.24 中,在卷积神经网络的前几层中,每一层的节点都被组织成一个三维矩阵。例如处理 CIFAR-10 数据集中的图片时,可以将输入层组织成一个 32×32×3 的三维矩阵。图 3.24 展示了卷积神经网络的一个连接示意图,从中可以看出卷积神经网络中前几层中每一个节点只和上一层中部分节点相连,一个卷积神经网络主要由 5 种结构组成。

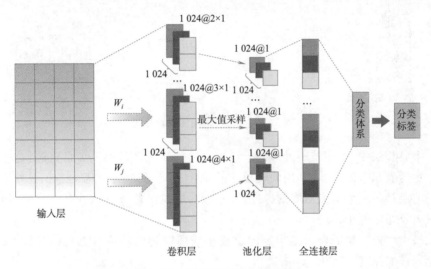

图 3.24 卷积神经网络基本结构

① 输入层。整个神经网络的输入,在处理图像的卷积神经网络中,它一般代表了一张图片的像素矩阵。通常用三维矩阵表示,三维矩阵的长和宽代表了图像的大小,而三维矩阵的深度代表了图像的色彩通道。例如黑白图片的深度为 1,而在 RGB 色彩模式下,图像的深度为 3。从输入层开始,卷积神经网络通过不同的神经网络结构将上一层的三维矩阵转化为下一层的三维矩阵,直到最后的全连接层。

② 卷积层。卷积层是一个卷积神经网络中最重要的部分。和传统全连接层不同,卷积层中每一个节点的输入只是上一层神经网络的一小块,这个小块常用的大小有 3×3 或者 5×5。卷积层试图将神经网络中的每一小块进行更加深入的分析,从而得到抽象程度更高的特征。一般来说,通过卷积层处理过的节点矩阵会变得更深。

③ 池化层(Pooling)。不会改变三维矩阵的深度,但是池化层可以缩小矩阵的大小。池化操作可以认为是将一张分辨率较高的图片转化为分辨率较低的图片。通过池化层,可以进一步缩小最后全连接层中节点的个数,从而达到减少整个神经网络中参数的目的。

④ 全连接层。在经过多轮卷积层和池化层的处理之后,在卷积神经网络的最后一般

会是由 1~2 个全连接层给出最后的分类结果。经过几轮卷积层和池化层的处理之后,可以认为图像中的信息已经被抽象成了信息含量更高的特征。可以将卷积层和池化层看成自动图像特征提取的过程。在特征提取完成之后,仍然需要使用全连接层完成分类任务。

⑤ 分类体系。主要用于分类问题,如使用 sofmax,可以得到当前样例属于不同种类的概率分布情况。

2. 卷积层

图 3.25 显示了 CNN 中最重要的部分,这部分称为卷积核(kernel),又称过滤器(filter)。卷积核可以将当前神经网络结构上的一个子节点矩阵转化为下一层神经网络上的一个单位节点矩阵,这里单位节点矩阵指的是长宽均为 1,深度不限的节点矩阵。

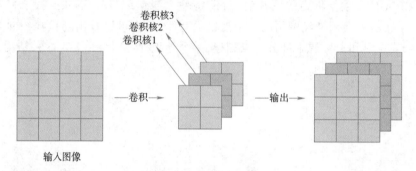

图 3.25 输入/输出与卷积核

在一个卷积层中,卷积核所处理的节点矩阵的长和宽都是由人工指定的,这个节点矩阵的尺寸称为卷积核的尺寸,常用的卷积核尺寸有 3×3 或 5×5。因为卷积核处理的矩阵深度和当前层神经网络节点矩阵的深度是一致的,所以虽然节点矩阵是三维的,但卷积核的尺寸只需要指定两个维度。

卷积核中另外一个需要人工指定的设置是处理得到的单位节点矩阵的深度,这个设置称为卷积核的深度。

注意卷积核的尺寸指的是一个卷积核输入节点矩阵的大小,而深度指的是输出单位节点矩阵的深度。为了直观地解释卷积核的前向传播过程,在下面的篇幅中将给出一个具体的样例,如图 3.26 所示。首先,暂时忽略第三维的通道维度,看看如何处理二维图像数据和卷积表示。假设输入图像是高度为 3、宽度为 3 的二维张量(即形状为 3×3)。卷积核的高度和宽度都是 2,而卷积核窗口(或卷积窗口)的形状由内核的高度和宽度决定(即 2×2)。

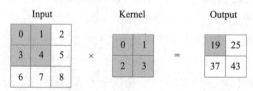

图 3.26 卷积计算

在卷积运算中,卷积窗口从输入张量的左上角开始,从左到右、从上到下滑动。当卷积窗口滑动到新位置时,包含在该窗口中的部分张量与卷积核张量进行按元素相乘,得到的张量再求和得到一个单一的标量值,由此可以得出这一位置的输出张量值。在如上例

子中,输出张量的四个元素由二维运算得到,这个输出高度为 2、宽度为 2,如下所示:

$$0 \times 0 + 1 \times 1 + 3 \times 2 + 4 \times 3 = 19$$
$$1 \times 0 + 2 \times 1 + 4 \times 2 + 5 \times 3 = 25$$
$$3 \times 0 + 4 \times 1 + 6 \times 2 + 7 \times 3 = 37$$
$$4 \times 0 + 5 \times 1 + 7 \times 2 + 8 \times 3 = 43$$

上例介绍了卷积层中计算一个卷积的前向传播过程。卷积层的前向传播过程就是将一个卷积核从神经网络当前层的左上角移动到右下角,并且在移动中计算每一个对应的单位矩阵得到的。

在这个过程中会涉及当前矩阵尺寸(长或宽) in_{length}、零填充 padding 值 p、卷积核尺寸(长或宽) f、卷积核移动的步长 stride 值 s 以及得到的矩阵尺寸(长或宽) out_{length} 等参数,它们的关系如下:

$$out_{length} = \frac{in_{length} + 2p - f}{s} + 1$$

在 CNN 中,每一个卷积层自己内部使用的卷积核的参数是一致的,这是 CNN 一个非常重要的特质。直观上理解,这样共享卷积核的参数可以使得图像上的内容不受位置的影响。参数共享机制可以巨幅减少神经网络的参数,下面以输入层为 $32 \times 32 \times 3$,目标层为 $28 \times 28 \times 16$ 矩阵为例,来说明参数计算:

- 全连接,则参数为 $(32 \times 32 \times 3 + 1) \times 28 \times 28 \times 16 = 38\,547\,712$。
- CNN,取卷积核大小为 5×5,步长为 1, no padding,则参数为 $(5 \times 5 \times 3 + 1) \times 16 = 1\,216$。
- CNN 但参数不共享,取卷积核特征同上,则不同卷积核数目为 28×28,参数为 $(5 \times 5 \times 3 + 1) \times 16 \times 28 \times 28 = 953\,344$。

可以看到,CNN 的稀疏连接和权值共享,二者都可以极大地减少神经网络的参数。而且卷积层的参数个数和图片的大小无关,只和卷积核的尺寸、深度以及当前节点矩阵的深度有关。这使得 CNN 可以很好地扩展到更大的图像数据上。

TensorFlow 对 CNN 提供了很好的支持,以下程序实现了一个卷积层的前向传播过程:

```
#通过 tf.get_variable 方式创建卷积核的权重变量和偏置项变量。上面介绍了卷积层的参数个
#数只和卷积核的尺寸、深度以及当前层节点矩阵的深度有关,所以这里声明的参数变量是一个四维矩
#阵,前面两个维度代表了卷积核的尺寸,第三个维度表示当前的深度,第四个维度表示卷积核的深度
    filter_weights = tf.get_variable('weights', [5, 5, 3, 16], initializer =
tf.truncated_normal_initializer(stddev = 0.1))

#和卷积层的权重类似,当前层矩阵上不同位置的偏置项也是共享的,所以总共有下一层深度个不同
#偏置项,这里为 16
    biases = tf.get_variable('biases', [16], initializer = tf.constant_initializer(0.1))

#tf.nn.conv2d 提供了一个非常方便的函数来实现卷积层前向传播的算法。第一个参数为当前层的
#节点矩阵。注意这个矩阵是一个四维矩阵,后面三个维度对应一个节点矩阵,第一维对应一个输入 batch。
#比如在输入层,input[0,:,:,:]表示第一张图片,input[1,:,:,:]表示第二张图片,依此类推。第二
#个参数提供了卷积层的权重。第三个参数为不同维度上的步长。虽然第三个参数提供的是一个长度为
#4 的数组,但是第一维和最后一维的数字要求一定是 1,这是因为卷积层的步长只对矩阵的长和宽有效。
#最后的参数是填充(padding)的方法,TensorFlow 中提供 SAME 或是 VALID 两种选择,其中 VALID
#表示不添加,SAME 表示添加 0 填充(TensorFlow 中全 0 填充优先填充右下方)
```

```
conv = tf.nn.conv2d(input,filter_weight,strides = [1,1,1,1],padding = "SAME")

#tf.nn.bias_add可以给每一个节点加上偏置项。注意这里不能直接使用加法,因为矩阵上不同
#位置上的节点都需要加上同样的偏置项
bias = tf.nn.bias_add(conv,biases)

#将计算结果通过ReLU激活函数完成去线性化
actived_conv = tf.nn.relu(bias)
```

3. 池化层

在卷积层之间往往会加上一个池化层(pooling layer)。池化层可以非常有效地缩小矩阵的尺寸(池化层主要用于减小矩阵的长和宽。虽然池化层也可以减小矩阵深度,但是实践中一般不会这样使用),减少最后全连接层中的参数。使用池化层既可以加快计算速度也可以防止过拟合问题。

和卷积层类似,池化层前向传播的过程也是通过移动一个类似卷积核的结构完成的(本书中称为池化窗口)。不过池化窗口节点的加权和采用更加简单的最大值或者平均值运算:

- 使用最大值操作的池化层称为最大池化层(max pooling),这是被使用得最多的池化层结构。
- 使用平均值操作的池化层称为平均池化层(average pooling)。
- 其他池化层在实践中使用得比较少,本书不做过多介绍。

与卷积层的卷积核类似,池化层的池化窗口也需要人工设定池化窗口的尺寸、是否使用全0填充以及池化窗口移动的步长等设置,而且这些设置的意义也是一样的。卷积层和池化层中池化窗口移动的方式是相似的,唯一的区别在于卷积层使用的卷积核是横跨整个深度的(三维的),而池化层使用的池化窗口只影响一个深度上的节点(二维的)。所以池化层的池化窗口除了在长和宽两个维度移动,它还需要在深度维度移动。图3.27展示了一个最大池化层前向传播计算过程。

图3.27 池化计算

池化窗口从输入张量的左上角开始,从左到右、从上到下地在输入张量内滑动。在池化窗口到达的每个位置,它计算该窗口中输入子张量的最大值:

$$\max(0,1,3,4) = 4$$
$$\max(1,2,4,5) = 5$$
$$\max(3,4,6,7) = 7$$
$$\max(4,5,7,8) = 8$$

以下TensorFlow程序实现了最大池化层的前向传播算法:

```
#tf.nn.max_pool实现了最大池化层的前向传播过程,它的参数和tf.nn.conv2d函数类似。
#ksize提供了池化窗口的尺寸,strides提供了步长信息,padding提供了是否使用全0填充。
pool = tf.nn.maxpool (actived_conv, ksize = [1,3 ,3,1], strides = [1,2,2,1],
padding = "SAME")
```

对比池化层和卷积层前向传播在TensorFlow中的实现,可以发现函数的参数形式是相

似的。在 tf.nn.max_pool 函数中：
- 首先需要传入当前层的节点矩阵，该矩阵是一个四维矩阵，格式和 tf.nn.conv2d 函数中的第一个参数一致。
- 第二个参数为池化窗口的尺寸。虽然给出的是一个长度为 4 的一维数组，但是这个数组的第一个和最后一个数必须为 1。这意味着池化层的池化窗口是不可以跨不同输入样例或者节点矩阵深度的。在实际应用中使用得最多的池化层池化窗口尺寸为[1,2,2,1]或者[1,3,3,1]。
- 第三个参数为步长。它和 tf.nn.conv2d 函数中步长的意义是一样的，而且第一维和最后一维也只能为 1。这意味着在 TensorFlow 中，池化层不能减少输入样例的个数或者节点矩阵的深度。
- 最后一个参数指定了填充方式。该参数有两种取值——VALID 或者 SAME，其中 VALID 表示不使用全 0 填充，SAME 表示使用全 0 填充。

TensorFlow 还提供了 tf.nn.avg_pool 函数来实现平均池化层，其调用格式与 tf.nn.max_pool 函数是一致的。

单元任务 11　认识卷积和池化操作

假设输入矩阵：

$$M = \begin{pmatrix} 1 & -1 & 0 \\ -1 & 2 & 1 \\ 0 & 2 & -2 \end{pmatrix}$$

卷积矩阵：

$$W = \begin{pmatrix} 1 & -1 \\ 0 & 2 \end{pmatrix}$$

实现一个简单的卷积和池化运算过程。

```
import tensorflow as tf
import numpy as np

#定义输入矩阵
M = np.array(
    [[[1],[-1],[0]],
    [[-1],[2],[1]],
    [[0],[2],[-2]]])

#定义参数
filter_weight = tf.get_variable('weights',[2,2,1,1],initializer = tf.constant_initializer([[1,-1],[0,2]]))
biases = tf.get_variable('biases',[1],initializer = tf.constant_initializer(1))

#调整输入的格式符合 TensorFlow 的要求
M = np.asarray(M,dtype = 'float32')
M = M.reshape(1,3,3,1)
#计算矩阵通过卷积层过滤器和池化层过滤器计算后的结果
```

```
x = tf.placeholder('float32',[1,None,None,1])

conv = tf.nn.conv2d(x,filter_weight,strides = [1,2,2,1],padding = 'SAME')
bias = tf.nn.bias_add(conv,biases)
pool = tf.nn.avg_pool(x,ksize = [1,2,2,1],strides = [1,2,2,1],padding = 'SAME')

with tf.Session() as sess:
    tf.global_variables_initializer().run()
    convoluted_M = sess.run(bias,feed_dict = {x:M})
    pooled_M = sess.run(pool,feed_dict = {x:M})

    print("convoluted_M: \n",convoluted_M)
    print("pooled_M: \n",pooled_M)
```

输出结果为:

```
convoluted_M:
[[[[ 7.]
   [ 1.]]

  [[-1.]
   [-1.]]]]
pooled_M:
[[[[ 0.25]
   [ 0.5 ]]

  [[ 1.  ]
   [-2.  ]]]]
```

3.4.2 循环神经网络

1. 循环神经网络简介

循环神经网络(Recurrent Neural Network,RNN)源自1982年由Saratha Sathasivam提出的霍普菲尔德网络。霍普菲尔德网络因为实现困难,在提出时并没有得到很好的应用。该网络结构也于1986年后被全连接神经网络以及一些传统的机器学习算法所取代。

- 传统的机器学习算法非常依赖于人工提取的特征,使得基于传统机器学习的图像识别、语音识别以及自然语言处理等问题难以突破特征提取的瓶颈。
- 基于全连接神经网络的方法也存在参数太多、无法利用数据中的时间序列信息等问题。

随着更加有效的循环神经网络结构被不断提出,循环神经网络挖掘数据中的时序信息以及语义信息的深度表达能力被充分利用,并在语音识别、语言模型、机器翻译以及时序分析等方面实现了突破。

RNN的主要用途是处理和预测序列数据。在之前介绍的全连接神经网络或CNN模型中,网络结构都是从输入层到隐含层再到输出层,层与层之间是全连接或部分连接的,但每层之间的节点是无连接的。考虑这样一个问题,如果要预测句子的下一个单词是什么,一般需要用到当前单词以及前面的单词,因为句子中前后单词并不是独立的。例如,当前单词是"很",前一个单词是"天空",那么下一个单词很大概率是"蓝"。RNN 的来源就是为了刻画一个序列当前的输出与之前信息的关系。从网络结构上,RNN 会记忆之前的信息,并利用之前的信息影响后面结点的输出。也就是说,RNN 的隐藏层之间的结点是有连接的,隐藏层的输入不仅包括输入层的输出,还包括上一时刻隐藏层的输出。

2. 循环神经网络原理

在实际应用中,经常会遇到很多序列型的数据,例如:
- 自然语言处理问题。x_1 可以看作第一个单词,x_2 可以看作第二个单词,依此类推。
- 语音处理。此时,x_1、x_2、x_3……是每帧的声音信号。
- 时间序列问题。例如每天的股票价格等。

其单个序列如图 3.28 所示。

图 3.28 输入序列

注:图中的圆圈表示向量,箭头表示对向量做变换。
RNN 引入了隐状态 h(hidden state),h 可对序列数据提取特征,接着再转换为输出。
为了便于理解,先计算 h_1,如图 3.29 所示。

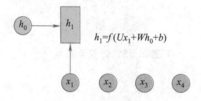

图 3.29 计算隐状态 h_1

RNN 中,每个步骤使用的参数 U,W,b 相同,h_2 的计算方式与 h_1 类似,其计算结果如图 3.30 所示。

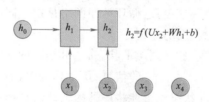

图 3.30 计算隐状态 h_2

以相似的方法计算 h_3、h_4,可得图 3.31。

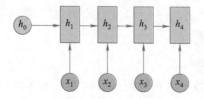

图 3.31 计算隐状态 h_3 和 h_4

接下来,计算 RNN 的输出 y_1,采用 Softmax 作为激活函数,根据 $y_n = f(Wx + b)$ 函数计算 y_1,如图 3.32 所示。

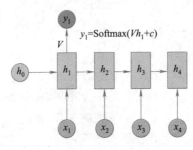

图 3.32　计算输出 y_1

使用和 y_1 相同的参数 V、c,得到 y_1、y_2、y_3、y_4 的输出结构,如图 3.33 所示。

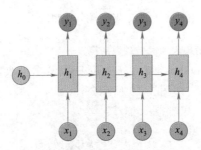

图 3.33　计算全部输出

以上即为最经典的 RNN 结构,其输入为 x_1、x_2、x_3、x_4,输出为 y_1、y_2、y_3、y_4,当然实际中最大值为 y_n,这里为了便于理解和展示,只计算 4 个输入和输出。从以上结构可看出,RNN 结构的输入和输出等长。

在每一个时刻,RNN 的模块 A 在读取了 x_t 和 h_{t-1} 之后会生成新的隐状态 h_t,并产生本时刻的输出,RNN 网络理论上可以看作同一神经网络结构被无限复制的结果。正如卷积神经网络在不同的空间位置共享参数,循环神经网络是在不同时间位置共享参数,从而能够使用有限的参数处理任意长度的序列。

RNN 当前的状态 h_t,是根据上一时刻的状态 h_{t-1} 和当前的输入 x_t 共同决定的。在时刻 t,状态 h_{t-1} 浓缩了前面序列 $x_0, x_1, \cdots, x_{t-1}$ 的信息,用于作为输出 o_t 的参考。由于序列的长度可以无限延长,维度有限的 h 状态不可能将序列的全部信息都保存下来,因此模型必须学习只保留与后面任务 $y_t, y_{t+1}, \cdots$ 相关的最重要的信息。

RNN 对长度为 N 的序列展开之后,可以视为一个有 N 个中间层的前馈神经网络。这个前馈神经网络没有循环连接,因此可以直接使用反向传播算法进行训练,而不需要任何特别的优化算法。这样的训练方法称为"沿时间反向传播"(Back-Propagation Through Time),是训练循环神经网络最常见的方法。

3. 循环神经网络应用场景

从 RNN 的结构特征容易看出它最擅长解决与时间序列相关的问题。RNN 也是处理这类问题时最自然的神经网络结构。对于一个序列数据,可以将该序列上不同时刻的数据依次传入 RNN 的输入层,而输出可以是对序列中下一个时刻的预测,也可以是对当前时刻信息的处理结果(比如语音识别结果)。循环神经网络要求每一个时刻都有一个输入,但是不一定每个时刻都需要有输出。在过去几年中,RNN 已经被广泛地应用在语音识别、语言模型、机器翻译以及时序分析等问题上,并取得了巨大的成功。

下面以机器翻译为例来介绍 RNN 是如何解决实际问题的。RNN 中每一个时刻的输入为需要翻译的句子中的单词。如图 3.34 所示,需要翻译的句子为 ABCD,那么 RNN 第一段每一个时刻的输入就分别是 A、B、C 和 D,然后用作待翻译句子的结束符。在第一段中,循环神经网络没有输出。从结束符开始,循环神经网络进入翻译阶段。该阶段中每一个时刻的输入是上一个时刻的输出,而最终得到的输出就是句子 ABCD 翻译的结果。

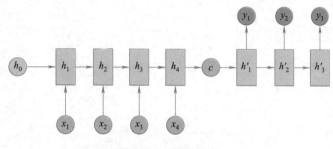

图 3.34　RNN 结构

4. 循环神经网络前向传播

RNN 中的状态是通过一个向量来表示的,这个向量的维度又称 RNN 隐藏层的大小,假设其为 n。循环体中神经网络的输入有两部分,一部分为上一时刻的状态,另一部分为当前时刻的输入样本。对于时间序列数据来说(比如不同时刻商品的销量),每一时刻的输入样例可以是当前时刻的数值;对于文本数据来说,输入样例可以是当前单词对应的单词向量(word embedding)。

假设输入向量的维度为 x,隐藏状态的维度为 n,那么图中循环体的全连接层神经网络的输入大小为 $n+x$。也就是将上一时刻的状态与当前时刻的输入拼接成一个大的向量作为循环体中神经网络的输入。因为该全连接层的输出为当前时刻的状态,于是输出层的节点个数也为 n,循环体中的参数个数为 $(n+x)\times n+n$ 个。从图 3.25 中可以看到,循环体中的神经网络输出不但提供给了下一时刻作为状态,同时也会提供给当前时刻的输出。注意循环体状态与最终输出的维度通常不同,因此为了将当前时刻的状态转化为最终的输出,RNN 还需要另外一个全连接神经网络来完成这个过程。这和 CNN 中最后的全连接层的意义是一样的。类似的,不同时刻用于输出的全连接神经网络中的参数也是一致的。为了让读者对 RNN 的前向传播有一个更加直观的认识,图 3.26 展示了一个 RNN 前向传播的具体计算过程。

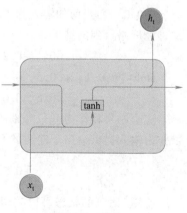

图 3.35　输入

在图 3.36 中,假设状态的维度为 2,输入、输出的维度都为 1,而且循环体中全连接层

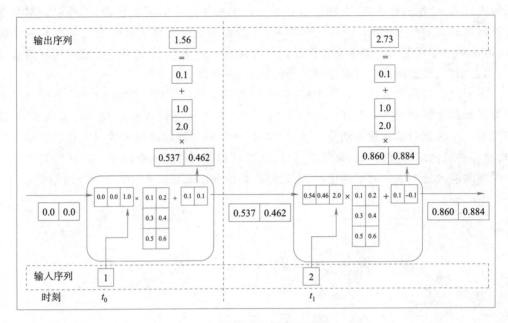

图 3.36 RNN 计算原理

中的权重为：

$$w_{rnn} = \begin{bmatrix} 0.1 & 0.2 \\ 0.3 & 0.4 \\ 0.5 & 0.6 \end{bmatrix}$$

偏置项的大小为 $b_{rnn} = [0.1, -0.1]$，用于输出的全连接层权重为：

$$w_{output} = \begin{bmatrix} 1.0 \\ 2.0 \end{bmatrix}$$

偏置项的大小为 $b_{output} = 0.1$。那么在时刻 t_0，因为没有上一时刻，所以将状态初始化为 $h_{init} = [0,0]$，而当前的输入为1，所以拼接得到的向量为 $[0,0,1]$，通过循环体中的全连接层神经网络得到的结果为：

$$\tanh\left([0,0,1] \times \begin{bmatrix} 0.1 & 0.2 \\ 0.3 & 0.4 \\ 0.5 & 0.6 \end{bmatrix} + [0.1, -0.1] \right) = \tanh([0.6, 0.5]) = [0.537, 0.462]$$

这个结果将作为下一时刻的输入状态，同时 RNN 也会使用该状态生成输出。将该向量作为输入提供给用于输出的全连接神经网络可以得到 t_0 时刻的最终输出：

$$[0.537, 0.462] \times \begin{bmatrix} 1.0 \\ 2.0 \end{bmatrix} + 0.1 = 1.56$$

使用 t_0 时刻的状态可以类似地推导得出 t_1 时刻的状态为 $[0.860, 0.884]$，而 t_1 时刻的输出为 2.73。在得到 RNN 的前向传播结果之后，可以和其他神经网络类似地定义损失函数。RNN 唯一的区别在于因为它每个时刻都有一个输出，所以 RNN 的总损失为所有时刻（或者部分时刻）上的损失函数的总和。

单元任务 12　循环神经网络前向传播

以下代码实现了这个简单的循环神经网络前向传播的过程：

```
import numpy as np

X = [1,2]
state = [0.0,0.0]
#分开定义不同输入部分的权重以方便操作
w_cell_state = np.asarray([[0.1,0.2],[0.3,0.4]])
w_cell_input = np.asarray([0.5,0.6])
b_cell = np.asarray([0.1,-0.1])

#定义用于输出的全连接层参数
w_output = np.asarray([[1.0],[2.0]])
b_output = 0.1

#执行前向传播过程
for i in range(len(X)):
    before_activation = np.dot(state,w_cell_state) + X[i]* w_cell_input + b_cell
    state = np.tanh(before_activation)
    final_output = np.dot(state,w_output) + b_output
    print("before activation: ",before_activation)
    print("state: ",state)
    print("output: ",final_output)
```

输出结果为：

```
before activation:  [0.6 0.5]
state:  [0.53704957 0.46211716]
output:  [1.56128388]
before activation:  [1.2923401  1.39225678]
state:  [0.85973818 0.88366641]
output:  [2.72707101]
```

和其他神经网络类似，在定义完损失函数之后，套用 TensorFlow 就可以自动完成模型训练的过程。这里唯一需要特别指出的是，理论上 RNN 可以支持任意长度的序列，然而在实际训练过程中，如果序列过长：

- 一方面会导致优化时出现梯度消散和梯度爆炸的问题。
- 另一方面，展开后的前馈神经网络会占用过大的内存。

所以实际中一般会规定一个最大长度，当序列长度超过规定长度之后会对序列进行截断。

3.4.3　长短期记忆

1. 长短期记忆网络简介

RNN 通过保存历史信息来帮助当前的决策，例如使用之前出现的单词来加强对当前文字的理解。RNN 可以更好地利用传统神经网络结构所不能建模的信息，但同时，这也带来了更大的技术挑战——长期依赖（long-term dependencies）问题。在有些问题中，模型仅仅需要短期内的信息来执行当前的任务。例如，预测短语"大海的颜色是蓝色"中的最后一个单词"蓝色"时，模型并不需要记忆这个短语之前更长的上下文信息，因为这一句话已经包含了足够的信息来预测最后一个词。在这样的场景中，相关的信息和待预测的词的位置之间的间隔很小，RNN 可以比较容易地利用先前信息。

但同样也会有一些上下文场景更加复杂的情况。例如当模型试着去预测段落"某地开设了大量工厂，空气污染十分严重……这里的天空都是灰色的"的最后一个单词时，仅仅根据短期依赖就无法很好地解决这种问题。因为只根据最后一小段，最后一个词可以是"蓝色的"或者"灰色的"。但如果模型需要预测清楚具体是什么颜色，就需要考虑先前提到但离当前位置较远的上下文信息。因此，当前预测位置和相关信息之间的文本间隔就有可能变得很大。当这个间隔不断增大时，简单 RNN 有可能会丧失学习到距离如此远的信息的能力。或者在复杂语言场景中，有用信息的间隔有大有小、长短不一，循环神经网络的性能也会受到限制。

为了解决该问题，研究人员提出了许多解决办法，其中应用最成功的就是门限 RNN (Gated RNN)，而长短期记忆网络(Long Short-Term Memory, LSTM)就是门限 RNN 中最常用的一种。有漏单元通过设计连接间的权重系数，从而允许 RNN 累积距离较远节点间的长期联系；而门限 RNN 则泛化了这样的思想，允许在不同时刻改变该系数，且允许网络忘记当前已经累积的信息。

2. 长短期记忆网络原理

在很多任务中，采用 LSTM 结构的 RNN 比标准的 RNN 表现更好。LSTM 结构是由 Sepp Hochreiter 和 Jurgen Schmidhuber 于 1997 年提出的，它是一种特殊的循环体结构。如图 3.37 所示，与单一 tanh 循环体结构不同，LSTM 是一种拥有三个"门"结构的特殊网络结构。

"门结构"：LSTM 靠一些"门"的结构让信息有选择性地影响循环神经网络中每个时刻的状态。所谓"门"的结构就是一个使用 sigmoid 神经网络和一个按位做乘法的操作，这两个操作合在一起就是一个"门"的结构。之所以该结构称为"门"是因为使用 sigmoid 作为激活函数的全连接神经网络层会输出一个 0~1 之间的数值，描述当前输入有多少信息量可以通过该结构。于是这个结构的功能就类似于一扇门，当门打开时(sigmoid 神经网络层输出为 1 时)，全部信息都可以通过；当门关上时(sigmoid 神经网络层输出为 0 时)，任何信息都无法通过。LSTM 拥有三个门，分别是遗忘门、输入门和输出门，本节下面的篇幅将介绍每一个"门"是如何工作的。

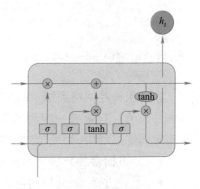

图 3.37 LSTM 门结构

- "遗忘门"的作用是让循环神经网络"忘记"之前没有用的信息。例如，一段文章中先介绍了某地原来是绿水蓝天，但后来被污染了。于是在看到被污染了之后，循环神经网络应该"忘记"之前绿水蓝天的状态。这个工作是通过"遗忘门"完成的。"遗忘门"会根据当前的输入 x_t 和上一时刻的输出 h_{t-1} 决定哪一部分记忆需要被遗忘。假设状态 c 的维度为 n。"遗忘门"会根据当前的输入 x_t 和上一时刻的输

出 h_{t-1} 计算一个维度为 n 的向量,它在每一维度上的值都在 $(0,1)$ 范围内。再将上一时刻的状态 c_{t-1} 与 f 向量按位相乘,那么 f 取值接近 0 的维度上的信息就会被"忘记",而 f 取值接近 1 的维度上的信息会被保留,如图 3.38 所示。

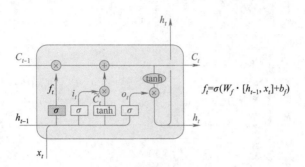

图 3.38　遗忘门

- 在 RNN "忘记"了部分之前的状态后,它还需要从当前的输入补充最新的记忆,这个过程就是"输入门"完成的。如图 3.39 和图 3.40 所示,"输入门"会根据 x_t 和 h_{t-1} 决定哪些信息加入到状态 c_{t-1} 中生成新的状态 c_t。例如,当看到文章中提到环境被污染之后,模型需要将这个信息写入新的状态。这时"输入门"和需要写入的新状态都从 x_t 和 h_{t-1} 计算产生。通过"遗忘门"和"输入门",LSTM 结构可以更加有效地决定哪些信息应该被遗忘,哪些信息应该得到保留。

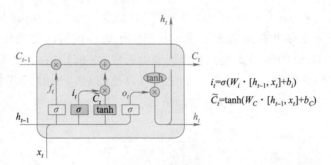

图 3.39　输入门

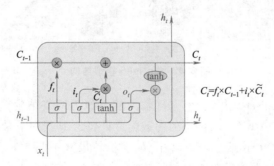

图 3.40　输入门

- "输出门":LSTM 结构在计算得到新的状态 c_t 后需要产生当前时刻的输出,这个过程是通过"输出门"完成的,如图 3.41 所示。"输出门"会根据最新的状态c_t、上一时刻的输出h_{t-1}和当前的输入x_t决定该时刻的输出值。例如,当前的状态为被污染,那么"天空的颜色"后面的单词很可能是"灰色的"。

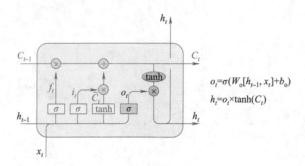

图 3.41 输出门

3.4.4 常见卷积神经主干网络

在前面章节介绍了 CNN 特有的两种网络结构——卷积层和池化层,但是怎么去组合它们更有可能解决真实的图像处理问题呢?这一节将介绍一些经典的 CNN 结构,通过这些经典的 CNN 网络结构可以总结出 CNN 结构设计的一些模式。

1. LeNet

LeNet-5 模型是 Yann LeCun 教授于 1998 年在论文 *Gradient based learning applied to document recognition* 中提出的,它是第一个成功应用于数字识别问题的卷积神经网络。其命名来源于作者 LeCun 的名字,5 则是其研究成果的代号,在 LeNet-5 之前还有 LeNet-4 和 LeNet-1,它们在 MNIST 数据集上,LeNet-5 模型可以达到 99.2% 的正确率。LeNet-5 模型总共有 7 层,图 3.42 展示了 LeNet-5 模型的结构。

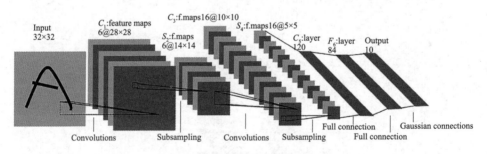

图 3.42 LeNet-5 模型结构

如图 3.42 所示,LeNet-5 共包含 7 层(输入层不作为网络结构),分别由 2 个卷积层、2 个下采样层和 3 个连接层组成,网络的参数配置见表 3.2,其中下采样层和全连接层的核尺寸分别代表采样范围和连接矩阵的尺寸(如卷积核尺寸中的 5×5×1/1,6 表示核大小为 5×5×1、步长为 1 且核个数为 6 的卷积核)。

表 3.2　LeNet-5 网络结构

网 络 层	输入尺寸	核 尺 寸	输出尺寸	可训练参数量
卷积层 C_1	$32 \times 32 \times 1$	$5 \times 5 \times 1/1,6$	$28 \times 28 \times 6$	$(5 \times 5 \times 1 + 1) \times 6$
下采样层 S_2	$28 \times 28 \times 6$	$2 \times 2/2$	$14 \times 14 \times 6$	$(1+1) \times 6$
卷积层 C_3	$14 \times 14 \times 6$	$5 \times 5 \times 6/1,16$	$10 \times 10 \times 16$	1516
下采样层 S_4	$10 \times 10 \times 16$	$2 \times 2/2$	$5 \times 5 \times 16$	$(1+1) \times 16$
卷积层 C_5	$5 \times 5 \times 16$	$5 \times 5 \times 16/1,120$	$1 \times 1 \times 120$	$(5 \times 5 \times 16 + 1) \times 120$
全连接层 F_6	$1 \times 1 \times 120$	120×84	$1 \times 1 \times 84$	$(120+1) \times 84$
输出层	$1 \times 1 \times 84$	84×10	$1 \times 1 \times 10$	$(84+1) \times 10$

LeNet 网络的特点如下：

- 卷积网络使用一个 3 层的序列组合：卷积、下采样（池化）、非线性映射（LeNet-5 最重要的特性，奠定了目前深层卷积网络的基础）。
- 使用卷积提取空间特征。
- 使用映射的空间均值进行下采样。
- 使用 tanh 或 sigmoid 进行非线性映射。
- 多层神经网络（MLP）作为最终的分类器。
- 层间的稀疏连接矩阵以避免巨大的计算开销。

2. AlexNet

在 LeNet 提出后，卷积神经网络在计算机视觉和机器学习领域中很有名气。但卷积神经网络并没有主导这些领域。这是因为虽然 LeNet 在小数据集上取得了很好的效果，但是在更大、更真实的数据集上训练卷积神经网络的性能和可行性还有待研究。事实上，在 20 世纪 90 年代初到 2012 年之间的大部分时间里，神经网络往往被其他机器学习方法超越，如支持向量机（support vector machines）。

2012 年，AlexNet 横空出世。它首次证明了学习到的特征可以超越手工设计的特征。它一举打破了计算机视觉研究的现状。AlexNet 使用了 8 层卷积神经网络，并以很大的优势赢得了 2012 年 ImageNet 图像识别挑战赛。

AlexNet 和 LeNet 的设计理念非常相似，但也存在显著差异。首先，AlexNet 比相对较小的 LeNet5 要深得多。AlexNet 由 8 层组成：5 个卷积层、2 个全连接隐藏层和 1 个全连接输出层。其次，AlexNet 使用 ReLU 而不是 sigmoid 作为其激活函数。下面，深入研究 AlexNet 的细节。

如图 3.43 所示，除去下采样层（池化层）和局部响应规范化操作，AlexNet 一共包含 8 层，前 5 层由卷积层组成，而剩下的 3 层为全连接层。网络结构分为上下两层，分别对应两个 GPU 的操作过程，除了中间某些层（C_3 卷积层和 $F_{6～8}$ 全连接层会有 GPU 间的交互），其他层两个 GPU 分别计算结果。最后一层全连接层的输出作为 softmax 的输入，得到 1 000 个图像分类标签对应的概率值。除去 GPU 并行结构的设计，AlexNet 网络结构与 LeNet 十分相似，其网络的参数配置见表 3.3。

在 AlexNet 的第一层，卷积窗口的形状是 11×11。由于 ImageNet 中大多数图像的宽和高比 MNIST 图像的多 10 倍以上，因此，需要一个更大的卷积窗口来捕获目标。第二层中

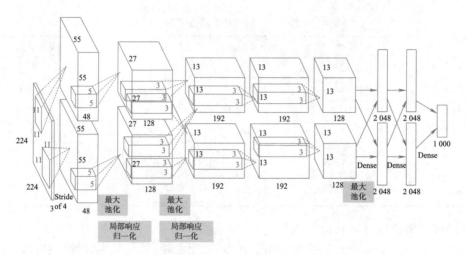

图 3.43 AlexNet 网络结构

的卷积窗口形状被缩减为 5×5,然后是 3×3。此外,在第一层、第二层和第五层卷积层之后,加入窗口形状为 3×3、步幅为 2 的最大池化层。而且,AlexNet 的卷积通道数目是 LeNet 的 10 倍。

表 3.3 AlexNet 网络结构

网 络 层	输入尺寸	核 尺 寸	输出尺寸	可训练参数量
卷积层 C_1	$224 \times 224 \times 3$	$11 \times 11 \times 3/4, 48(\times 2_{GPU})$	$55 \times 55 \times 48(\times 2_{GPU})$	$(11 \times 11 \times 3 + 1) \times 48 \times 2$
下采样层 S_{max}	$55 \times 55 \times 48(\times 2_{GPU})$	$3 \times 3/2(\times 2_{GPU})$	$27 \times 27 \times 48(\times 2_{GPU})$	0
卷积层 C_2	$27 \times 27 \times 48(\times 2_{GPU})$	$5 \times 5 \times 48/1, 128(\times 2_{GPU})$	$27 \times 27 \times 128(\times 2_{GPU})$	$(5 \times 5 \times 48 + 1) \times 128 \times 2$
下采样层 S_{max}	$27 \times 27 \times 128(\times 2_{GPU})$	$3 \times 3/2(\times 2_{GPU})$	$13 \times 13 \times 128(\times 2_{GPU})$	0
卷积层 C_3	$13 \times 13 \times 128 \times 2_{GPU}$	$3 \times 3 \times 256/1, 192(\times 2_{GPU})$	$13 \times 13 \times 192(\times 2_{GPU})$	$(3 \times 3 \times 256 + 1) \times 192 \times 2$
卷积层 C_4	$13 \times 13 \times 192(\times 2_{GPU})$	$3 \times 3 \times 192/1, 192(\times 2_{GPU})$	$13 \times 13 \times 192(\times 2_{GPU})$	$(3 \times 3 \times 192 + 1) \times 192 \times 2$
卷积层 C_5	$13 \times 13 \times 192(\times 2_{GPU})$	$3 \times 3 \times 192/1, 128(\times 2_{GPU})$	$13 \times 13 \times 128(\times 2_{GPU})$	$(3 \times 3 \times 192 + 1) \times 128 \times 2$
下采样层 S_{max}	$13 \times 13 \times 128(\times 2_{GPU})$	$3 \times 3/2(\times 2_{GPU})$	$6 \times 6 \times 128(\times 2_{GPU})$	0
全连接层 F_6	$6 \times 6 \times 128 \times 2_{GPU}$	$9\,216 \times 2\,048(\times 2_{GPU})$	$1 \times 1 \times 2\,048(\times 2_{GPU})$	$(9\,216 + 1) \times 2\,048 \times 2$
全连接层 F_7	$1 \times 1 \times 2\,048 \times 2_{GPU}$	$4\,096 \times 2\,048(\times 2_{GPU})$	$1 \times 1 \times 2\,048(\times 2_{GPU})$	$(4\,096 + 1) \times 2\,048 \times 2$
全连接层 F_8	$1 \times 1 \times 2\,048 \times 2_{GPU}$	$4\,096 \times 1\,000$	$1 \times 1 \times 1\,000$	$(4\,096 + 1) \times 1\,000 \times 2$

AlexNet 网络的特点:
- 所有卷积层都使用 ReLU 作为非线性映射函数,使模型收敛速度更快。
- 在多个 GPU 上进行模型的训练,不但可以提高模型的训练速度,还能提升数据的使用规模。
- 使用 LRN 对局部的特征进行归一化,结果作为 ReLU 激活函数的输入能有效降低错误率。
- 重叠最大池化(overlapping max pooling),即池化范围 z 与步长 s 存在关系 $z > s$(如 S_{max} 中核尺度为 $3 \times 3/2$),避免平均池化(average pooling)的平均效应。
- 使用随机丢弃技术(dropout)选择性地忽略训练中的单个神经元,避免模型的过拟合。

3. Network In Network

Network In Network(NIN)是由 MinLin 等人提出,在 CIFAR-10 和 CIFAR-100 分类任务中达到当时的最好水平,因其网络结构是由三个多层感知机堆叠而称为 NIN。NIN 以一种全新的角度审视了卷积神经网络中的卷积核设计,通过引入子网络结构代替纯卷积中的线性映射部分,这种形式的网络结构激发了更复杂的卷积神经网络的结构设计。

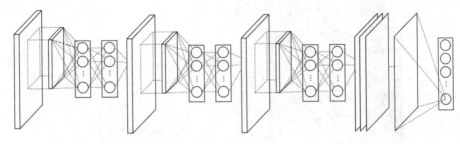

图 3.44　NIN 网络结构

NIN 由三层的多层感知卷积层构成,每一层多层感知卷积层内部由若干层的局部全连接层和非线性激活函数组成,代替了传统卷积层中采用的线性卷积核。在网络推理时,这个多层感知器会对输入特征图的局部特征进行划窗计算,并且每个划窗的局部特征图对应的乘积的权重是共享的,这两点是和传统卷积操作完全一致的,最大的不同在于多层感知器对局部特征进行了非线性的映射,而传统卷积的方式是线性的。NIN 的网络参数配置见表 3.4。

表 3.4　NIN 网络结构

网 络 层	输入尺寸	核 尺 寸	输出尺寸	参数个数
局部全连接层 L_{11}	$32 \times 32 \times 3$	$(3 \times 3) \times 16/1$	$30 \times 30 \times 16$	$(3 \times 3 \times 3 + 1) \times 16$
全连接层 L_{12}	$30 \times 30 \times 16$	16×16	$30 \times 30 \times 16$	$((16+1) \times 16)$
局部全连接层 L_{21}	$30 \times 30 \times 16$	$(3 \times 3) \times 64/1$	$28 \times 28 \times 64$	$(3 \times 3 \times 16 + 1) \times 64$
全连接层 L_{22}	$28 \times 28 \times 64$	64×64	$28 \times 28 \times 64$	$((64+1) \times 64)$
局部全连接层 L_{31}	$28 \times 28 \times 64$	$(3 \times 3) \times 100/1$	$26 \times 26 \times 100$	$(3 \times 3 \times 64 + 1) \times 100$
全连接层 L_{32}	$26 \times 26 \times 100$	100×100	$26 \times 26 \times 100$	$((100+1) \times 100)$
全局平均采样 GAP	$26 \times 26 \times 100$	$26 \times 26 \times 100/1$	$1 \times 1 \times 100$	0

NIN 网络的特点:
- 使用多层感知机结构代替卷积的滤波操作,不但有效减少卷积核数过多而导致的参数量暴涨问题,还能通过引入非线性映射提高模型对特征的抽象能力。
- 使用全局平均池化来代替最后一个全连接层,能够有效地减少参数数量(没有可训练参数),同时池化用到了整个特征图的信息,对空间信息的转换更加鲁棒,最后得到的输出结果可直接作为对应类别的置信度。

4. VGGNet

VGGNet 是由牛津大学视觉几何小组(Visual Geometry Group,VGG)提出的一种深层卷积网络结构,其以 7.32% 的错误率赢得了 2014 年 ILSVRC 分类任务的亚军和 25.32%

的错误率夺得定位任务的冠军,网络名称 VGGNet 取自该小组名缩写。VGGNet 是首批把图像分类的错误率降低到 10% 以内的模型,同时该网络所采用的 3×3 卷积核的思想是后来许多模型的基础。VGGNet 网络结构如图 3.45 所示。

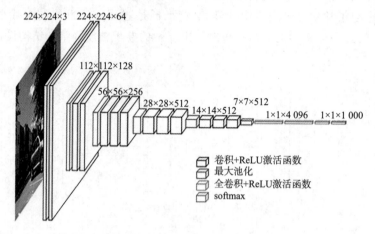

图 3.45　VGGNet 网络结构

原本的 VGGNet 包含了 6 个版本的演进,分别对应 VGG11、VGG11-LRN、VGG13、VGG16-1、VGG16-3 和 VGG19,不同的后缀数值表示不同的网络层数(VGG11-LRN 表示在第一层中采用了 LRN 的 VGG11;VGG16-1 表示后三组卷积块中最后一层卷积采用卷积核尺寸为 1×1;VGG16-3 表示卷积核尺寸为 3×3),本节介绍的 VGG16 为 VGG16-3。图的 VGG16 体现了 VGGNet 的核心思路,使用 3×3 的卷积组合代替大尺寸的卷积(2 个 3×3 卷积即可与 5×5 卷积拥有相同的感受野),网络参数设置见表 3.5。

表 3.5　VGGNet 网络结构

网 络 层	输入尺寸	核 尺 寸	输出尺寸	参数个数
卷积层 C_{11}	224×224×3	3×3 64/1	224×224×64	(3×3×3+1)×64
卷积层 C_{12}	224×224×64	3×3 64/1	224×224×64	(3×3×64+1)×64
下采样层 S_{max1}	224×224×64	2×2/2	112×112×64	0
卷积层 C_{21}	112×112×64	3×3 128/1	112×112×128	(3×3×64+1)×128
卷积层 C_{22}	112×112×128	3×3 128/1	112×112×128	(3×3×128+1)×128
下采样层 S_{max2}	112×112×128	2×2/2	56×56×128	0
卷积层 C_{31}	56×56×128	3×3 256/1	56×56×256	(3×3×128+1)×256
卷积层 C_{32}	56×56×256	3×3 256/1	56×56×256	(3×3×256+1)×256
卷积层 C_{33}	56×56×256	3×3 256/1	56×56×256	(3×3×256+1)×256
下采样层 S_{max3}	56×56×256	2×2/2	28×28×256	0
卷积层 C_{41}	28×28×256	3×3 512/1	28×28×512	(3×3×256+1)×512
卷积层 C_{42}	28×28×512	3×3 512/1	28×28×512	(3×3×512+1)×512
卷积层 C_{43}	28×28×512	3×3 512/1	28×28×512	(3×3×512+1)×512
下采样层 S_{max4}	28×28×512	2×2/2	14×14×512	0
卷积层 C_{51}	14×14×512	3×3 512/1	14×14×512	(3×3×512+1)×512
卷积层 C_{52}	14×14×512	3×3 512/1	14×14×512	(3×3×512+1)×512

续表

网络层	输入尺寸	核尺寸	输出尺寸	参数个数
卷积层C_{53}	$14 \times 14 \times 512$	$3 \times 3 \times 512/1$	$14 \times 14 \times 512$	$(3 \times 3 \times 512 + 1) \times 512$
下采样层S_{max5}	$14 \times 14 \times 512$	$2 \times 2/2$	$7 \times 7 \times 512$	0
全连接层FC_1	$7 \times 7 \times 512$	$(7 \times 7 \times 512) \times 4\,096$	$1 \times 4\,096$	$(7 \times 7 \times 512 + 1) \times 4\,096$
全连接层FC_2	$1 \times 4\,096$	$4\,096 \times 4\,096$	$1 \times 4\,096$	$(4\,096 + 1) \times 4\,096$
全连接层FC_3	$1 \times 4\,096$	$4\,096 \times 1\,000$	$1 \times 1\,000$	$(4\,096 + 1) \times 1\,000$

VGG 网络的特点:
- 整个网络都使用了同样大小的卷积核尺寸 3×3 和最大池化尺寸 2×2。
- 1×1 卷积的意义主要在于线性变换,而输入通道数和输出通道数不变,没有发生降维。
- 两个 3×3 的卷积层串联相当于 1 个 5×5 的卷积层,感受野大小为 5×5。同样地,3 个 3×3 的卷积层串联的效果则相当于 1 个 7×7 的卷积层。这样的连接方式使得网络参数量更小,而且多层的激活函数令网络对特征的学习能力更强。
- VGGNet 在训练时有一个小技巧,先训练浅层的简单网络 VGG11,再复用 VGG11 的权重来初始化 VGG13,如此反复训练并初始化 VGG19,能够使训练时收敛的速度更快。
- 在训练过程中使用多尺度的变换对原始数据做数据增强,使得模型不易过拟合

5. GoogLeNet

GoogLeNet 作为 2014 年 ILSVRC 在分类任务上的冠军,以 6.65% 的错误率力压 VGGNet 等模型,在分类的准确率上相比过去两届冠军都有很大的提升。从名字 GoogLeNet 可以知道这是来自谷歌工程师所设计的网络结构,而名字中 LeNet 更是致敬了 LeNet。GoogLeNet 中最核心的部分是其内部子网络结构 Inception,该结构的灵感来源于 NIN,至今已经历了四次版本迭代。GoogLeNet 网络结构如图 3.46 所示。

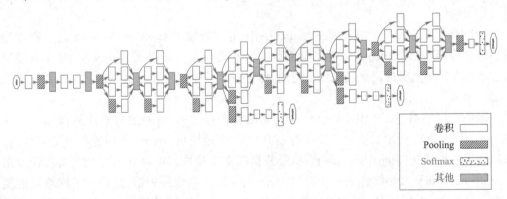

图 3.46 GoogLeNet 网络结构

如图 3.46 所示,GoogLeNet 相比于以前的卷积神经网络结构,除了在深度上进行了延伸,还对网络的宽度进行了扩展,整个网络由许多块状子网络堆叠而成,这个子网络构成了 Inception 结构。在同一层中采用不同的卷积核,并对卷积结果进行合并;Inception$_{v2}$ 组合不同卷积核的堆叠形式,并对卷积结果进行合并;Inception$_{v3}$ 则在 v_2 的基础上进行深度组合

的尝试；Inception$_{v4}$结构相比于前面的版本更加复杂，子网络中嵌套着子网络。GoogLeNet中Inception$_{v1}$网络参数配置见表3.6。

表3.6 GoogLeNet 的 Inception$_{v1}$ 网络结构

网 络 层	输入尺寸	核 尺 寸	输出尺寸	参数个数
卷积层C_{11}	$H \times W \times C_1$	$1 \times 1 \times C_2/2$	$\frac{H}{2} \times \frac{W}{2} \times C_2$	$(1 \times 1 \times C_1 + 1) \times C_2$
卷积层C_{21}	$H \times W \times C_2$	$1 \times 1 \times C_2/2$	$\frac{H}{2} \times \frac{W}{2} \times C_2$	$(1 \times 1 \times C_2 + 1) \times C_2$
卷积层C_{22}	$H \times W \times C_2$	$3 \times 3 \times C_2/1$	$H \times W \times C_2/1$	$(3 \times 3 \times C_2 + 1) \times C_2$
卷积层C_{31}	$H \times W \times C_1$	$1 \times 1 \times C_2/2$	$\frac{H}{2} \times \frac{W}{2} \times C_2$	$(1 \times 1 \times C_1 + 1) \times C_2$
卷积层C_{32}	$H \times W \times C_2$	$5 \times 5 \times C_2/1$	$H \times W \times C_2/1$	$(5 \times 5 \times C_2 + 1) \times C_2$
下采样层S_{41}	$H \times W \times C_1$	$3 \times 3/2$	$\frac{H}{2} \times \frac{W}{2} \times C_2$	0
卷积层C_{42}	$\frac{H}{2} \times \frac{W}{2} \times C_2$	$1 \times 1 \times C_2/1$	$\frac{H}{2} \times \frac{W}{2} \times C_2$	$(3 \times 3 \times C_2 + 1) \times C_2$
合并层M	$\frac{H}{2} \times \frac{W}{2} \times C_2 (\times 4)$	拼接	$\frac{H}{2} \times \frac{W}{2} \times (C_2 \times 4)$	0

GoogleNet 网络的特点：

- 采用不同大小的卷积核意味着不同大小的感受野，最后拼接意味着不同尺度特征的融合。
- 之所以卷积核大小采用1、3和5，主要是为了方便对齐。设定卷积步长 stride = 1 之后，只要分别设定 pad = 0、1、2，那么卷积之后便可以得到相同维度的特征，然后这些特征就可以直接拼接在一起了。
- 网络越到后面，特征越抽象，而且每个特征所涉及的感受野也更大了，因此随着层数的增加，3×3 和 5×5 卷积的比例也要增加。但是，使用 5×5 的卷积核仍然会带来巨大的计算量。为此，采用 1×1 卷积核进行降维。

6. ResNet

2015 年，ILSVRC 挑战赛的赢家 Kaiming He 等人开发的 Residual Network 是一种常用的网络结构，该网络的 top-5 错误率低到 3.6%，它使用了一个非常深的 CNN，由152 层组成。能够训练如此深的网络的关键是使用跳跃连接：一个层的输入信号也被添加到位于下一层的输出。

ResNet 网络参考了 VGG19 网络，在其基础上进行了修改，并通过短路机制加入了残差单元，如图3.47 所示。变化主要体现在 ResNet 直接使用 stride = 2 的卷积做下采样，并且用全局平均池化(global average pool)层替换了全连接层。ResNet 的一个重要设计原则是：当 feature map 大小降低到一半时，feature map 的数量增加一倍，这保持了网络层的复杂度。从图3.47 可知，ResNet 相比普通网络每两层间增加了短路机制，这就形成了残差学习，其中虚线表示 feature map 数量发生了改变。图3.47 展示的34-layer 的 ResNet，还可以构建更深的网络，见表3.7。从表中可以看到，对于 18-layer 和 34-layer 的 ResNet，其进行的两层间的残差学习，当网络更深时，其进行的是三层间的残差学习，三层卷积核分别是 1×1、3×3 和 1×1，值得注意的是隐含层的 feature map 数量比较小，是输出 feature map 数量的 1/4。

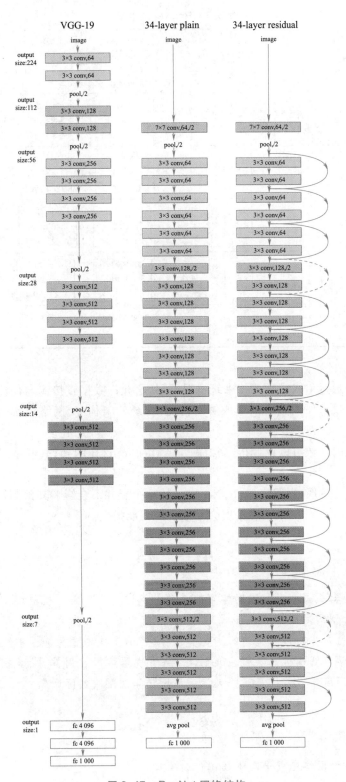

图 3.47　ResNet 网络结构

表 3.7 ResNet 网络架构

层名称	输出尺寸	18-layer	34-layer	50-layer	101-layer	152-layer
conv1	112×112	\multicolumn{5}{c}{7×7, 64, stride 2}				
conv2_x	56×56	$\begin{bmatrix}3\times3,64\\3\times3,64\end{bmatrix}\times2$	$\begin{bmatrix}3\times3,64\\3\times3,64\end{bmatrix}\times3$	$\begin{bmatrix}1\times1,64\\3\times3,64\\1\times1,256\end{bmatrix}\times3$	$\begin{bmatrix}1\times1,64\\3\times3,64\\1\times1,256\end{bmatrix}\times3$	$\begin{bmatrix}1\times1,64\\3\times3,64\\1\times1,256\end{bmatrix}\times3$
conv3_x	28×28	$\begin{bmatrix}3\times3,128\\3\times3,128\end{bmatrix}\times2$	$\begin{bmatrix}3\times3,128\\3\times3,128\end{bmatrix}\times4$	$\begin{bmatrix}1\times1,128\\3\times3,128\\1\times1,512\end{bmatrix}\times4$	$\begin{bmatrix}1\times1,128\\3\times3,128\\1\times1,512\end{bmatrix}\times4$	$\begin{bmatrix}1\times1,128\\3\times3,128\\1\times1,512\end{bmatrix}\times8$
conv4_x	14×4	$\begin{bmatrix}3\times3,256\\3\times3,256\end{bmatrix}\times2$	$\begin{bmatrix}3\times3,256\\3\times3,256\end{bmatrix}\times6$	$\begin{bmatrix}1\times1,256\\3\times3,256\\1\times1,1024\end{bmatrix}\times6$	$\begin{bmatrix}1\times1,256\\3\times3,256\\1\times1,1024\end{bmatrix}\times23$	$\begin{bmatrix}1\times1,256\\3\times3,256\\1\times1,1024\end{bmatrix}\times36$
conv5_x	7×7	$\begin{bmatrix}3\times3,512\\3\times3,512\end{bmatrix}\times2$	$\begin{bmatrix}3\times3,512\\3\times3,512\end{bmatrix}\times3$	$\begin{bmatrix}1\times1,512\\3\times3,512\\1\times1,2048\end{bmatrix}\times3$	$\begin{bmatrix}1\times1,512\\3\times3,512\\1\times1,2048\end{bmatrix}\times3$	$\begin{bmatrix}1\times1,512\\3\times3,512\\1\times1,2048\end{bmatrix}\times3$
	1×1	average pool, 1 000-d fc, softmax				
FLOPs		1.8×10^9	3.6×10^9	3.8×10^9	7.6×10^9	11.3×10^9

ResNet 网络的特点:
- ResNet 通过引入跳跃连接来绕过残差层,这允许数据直接流向任何后续层。这与传统的、顺序的 pipeline 形成鲜明对比:传统的架构中,网络依次处理低级 feature 到高级 feature。
- ResNet 的层数可以设计得非常深。而 ALexNet 这样的架构,网络层数属于浅层网络架构。
- 通过实验发现,训练好的 ResNet 中去掉单个层并不会影响其预测性能。而训练好的 AlexNet 等网络中,移除层会导致预测性能损失。

小 结

本单元主要对机器学习和深度学习的基础知识进行了讲解。机器学习是一门讨论各式各样的适用于不同问题的函数形式,以及如何使用数据来有效地获取函数参数具体值的学科;深度学习是指机器学习中的一类算法,它们的形式通常为多层神经网络。本单元四项任务:要求学生明确机器学习主要任务、熟悉机器学习常用算法、掌握深度学习相关知识基础,了解深度学习常用算法并灵活运用所学算法解决实际问题。

练 习

使用卷积神经网络完成手写体数字识别项目。

单元4　数据集和预处理

深度学习的发展,离不开数据。对于从事深度学习相关行业的工作者来说,必须对数据建立充分的认知。深度学习的训练往往需要海量的数据,而如今数据又是如此的宝贵,如何利用数据获得更好的效果呢?对于应用在深度学习中的数据,一方面需要寻找更多的数据,另一方面要对得到的数据进行预处理,提高数据的利用效率。

本单元系统地介绍深度学习数据的有关知识,主要包括两个方面:一个是数据集在深度学习中的发展与应用;另一个是图像数据的预处理方法。通过学习单元,读者可以了解到深度学习中常用的数据集,熟悉图像数据常用的预处理方法。本单元的知识导图如图4.1所示。

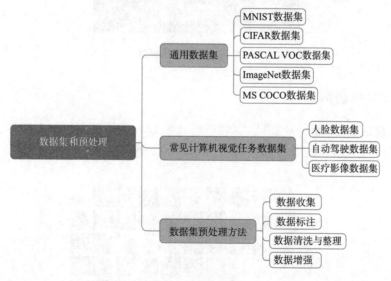

图4.1　知识导图

课程安排

课程任务	课程目标	安排课时
深度学习中的数据集	了解深度学习中常用的各类数据集及其特点	1
掌握数据预处理方法	了解数据收集方式、数据标注工具的使用方法,明确数据清洗与整理思路,熟练掌握常用数据增强方法	3

4.1 通用数据集

4.1.1 MNIST 数据集

MNIST 数据集是机器学习领域中非常经典的一个数据集,主要由一些手写数字的图片和相应的标签组成,图片一共有 10 类,分别对应从 0~9 共 10 个阿拉伯数字。由 60 000 个训练样本和 10 000 个测试样本组成,每个样本都是一张 28×28 像素的灰度手写数字图片,如图 4.2 所示。

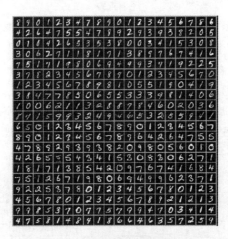

图 4.2 MNIST 数据集

4.1.2 CIFAR 数据集

CIFAR 数据集由国外一个团队收集整理,其中 CIFAR 数据集又根据所涉及的分类对象数量,可分为 CIFAR-10(见图 4.3)和 CIFAR-100。该数据集主要用于深度学习的图像分类,目前已被广泛应用。

图 4.3 CIFAR 数据集

常用的 CIFAR-10 数据集包含飞机、汽车、鸟等 10 个类别物体的 32×32 像素的彩色图片,每个类别有 6 000 张图片,计算可知,整个数据集共有 6 000×10＝60 000 张图片。数据集包含有训练集和测试集两个子集。训练集共 50 000 张图片,包含了来自 10 个类别的 1 000 张图片。测试集共 10 000 张图片,也包含了随机从每个类中抽取的 1 000 张图片。

4.1.3　PASCAL VOC 数据集

PASCAL VOC 数据集是目标检测经常用的一个数据集,自 2005 年起每年举办一次比赛,最开始只有 4 类,到 2007 年扩充为 20 个类,如图 4.4 所示,共有两个常用的版本:2007 和 2012。PASCAL VOC 数据集适用于目标检测、语义分割和实例分割等多个图像视觉任务。

图 4.4　PASCAL VOC 数据集

4.1.4　ImageNet 数据集

ImageNet 是一个计算机视觉系统识别项目,是目前世界上图像识别最大的数据库,如图 4.5 所示。ImageNet 是美国斯坦福的计算机科学家,模拟人类的识别系统建立的。能够通过图片识别物体。ImageNet 数据集文档详细有专门的团队维护,使用非常方便,在计算机视觉领域研究论文中应用非常广,几乎成了目前深度学习图像领域算法性能检验的"标准"数据集。ImageNet 数据集有 1 400 多万幅图片,涵盖 2 万多个类别;其中有超过百万的图片有明确的类别标注和图像中物体位置的标注。

ImageNet 是一项持续的研究工作,旨在为世界各地的研究人员提供易于访问的图像数据库。目前 ImageNet 中总共有 14 197 122 幅图像,总共分为 21 841 个类别。

4.1.5　MS COCO 数据集

MS COCO 数据集是微软团队发布的数据集,该数据集收集了大量包含常见物体的日常场景图片,如图 4.6 所示,并提供像素级的实例标注,以便精确地评估检测和分割算法的效果,致力于推动场景理解的研究进展。依托这一数据集,每年举办一次比赛,现已涵盖检测、分割、关键点识别、注释等机器视觉的中心任务，是继 ImageNet Chanllenge 以来最有影响力的竞赛之一。MS COCO 的检测任务共含有 80 个类，在 2014 年发布的数据规模分 train/val/test，分别为 80 000 张/40 000 张/40 000 张。

图 4.5　ImageNet 数据集

图 4.6　MS COCO 数据集

4.2 常见计算机视觉任务数据集

4.2.1 人脸数据集

1. PubFig 数据集

PubFig(Public Figures Face Database)数据集是哥伦比亚大学开放的公众人物脸部数据集,包含有200个人的58 000多张人脸图像,主要用于非限制场景下的人脸识别,如图4.7所示。

图 4.7　PubFig 数据集

2. CelebA 数据集

CelebA 数据集是香港中文大学汤晓鸥教授实验室公布的大型人脸识别数据集。包含有200 000张人脸图像,人脸属性有40多种,主要用于人脸属性的识别,如图4.8所示。

图 4.8　CelebA 数据集

3. LFW 数据集

LFW(Labeled Faces in the Wild Home)数据集是人脸识别项目常用的数据集,包含了来自1 680个人的13 000张人脸图像,如图4.9所示,数据是从网上搜集来的。

图 4.9　LFW 数据集

4.2.2 自动驾驶数据集

1. Kitti 数据集

Kitti 数据集由德国卡尔斯鲁厄理工学院和丰田美国技术研究院联合创办,是目前国际上最大的自动驾驶场景下的计算机视觉算法评测数据集。该数据集用于评测立体图像、光流、视觉测距、3D 物体检测和 3D 跟踪等计算机视觉技术在车载环境下的性能。Kitti 包含市区、乡村和高速公路等场景采集的真实图像数据,每张图像中最多达 15 辆车和 30 个行人,还有各种程度的遮挡与截断,如图 4.10 所示。

图 4.10 Kitti 数据集

2. Apollo 开源自动驾驶数据集

Apollo 开源自动驾驶数据集包括感知分类和路网数据等数十万帧逐像素语义分割标注的高分辨率图像数据,以及与其对应的逐像素语义标注,覆盖了来自三个城市的三个站点的地域。主要包含三部分:仿真数据集、演示数据集、标注数据集:

- 仿真数据集,包括自动驾驶虚拟场景和实际道路真实场景。
- 演示数据集,包括车载系统演示数据、标定演示数据、端到端演示数据、自定位模块演示数据。
- 标注数据集,包括 6 部分数据集:激光点云障碍物检测分类、红绿灯检测、Road Hackers、基于图像的障碍物检测分类、障碍物轨迹预测、场景解析。

图 4.11 Apollo 开源自动驾驶数据集

4.2.3 医疗影像数据集

1. 数字视网膜图像数据集

数字视网膜图像数据集主要用于视网膜图像中血管分割的比较研究,它由 40 张照片组成,其中 7 张显示出轻度早期糖尿病性视网膜病变的迹象。

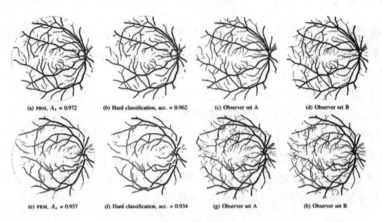

图 4.12 数字视网膜图像数据集

2. 皮肤损伤数据集

皮肤损伤数据集包含分类皮肤损伤的 23 000 张图像。含有恶性和良性的例子。每个示例均包含病变的图像,可进行有关病变的图像分类和图像分割任务的应用型研究。

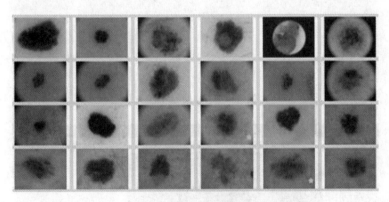

图 4.13 皮肤损伤数据集

4.3 数据集预处理方法

4.3.1 数据收集

在训练机器学习模型时,找到合适的数据集一直是个棘手的问题。数据采集是一个需要人力和物力的任务,通常数据采集可以根据具体的项目课题通过合理的手段收集。通常有自行制作数据集、寻找已有的公开数据集和数据爬虫等方式,但是通过这些方式获取的数据集常常存在一定的局限性,普遍存在数据集不规范、数据集缺失、数据样本不均衡等问题。

4.3.2 数据标注

数据标注是数据收集后的一个重要步骤,数据标注就是对未处理的初级数据,包括语音、图片、文本、视频等进行加工处理,转换为属性标签以训练数据集。根据不同的目标任务类型,标注任务也不尽相同:

- 分类标注:对图片进行分类。
- 检测标注:对图片中出现的物体检测其位置。
- 分割标注:对图片进行切割。

数据处理平台的数据标注工具是一款简单独立的手动注释应用,适用于图像和视频。通过 Web 浏览器运行,不需要任何安装或设置。支持标注的区域组件有:矩形、圆形、椭圆形、多边形、点和折线等,如图 4.14 和图 4.15 所示。

图 4.14 数据标注任务

图 4.15 数据标注区

4.3.3 数据清洗与整理

数据在采集完之后,往往包含着噪声、缺失数据、不规则数据等各种问题,因此需要对其进行清洗和整理工作,主要包括以下内容。

1. 数据规范化管理

规范化管理后的数据,才有可能成为一个标准的数据集,其中数据命名的统一是第一步。通常抓取和采集回来的数据没有统一、连续的命名,因此需要制定统一的格式,命名通常不要含有中文字符和不合法字符等,在后续使用过程中不能对数据集进行重命名,否则会造成数据无法回溯的问题,而导致数据丢失。另外,对于图像等数据,还需要统一格式,例如,把一批图片数据统一为 JPG 格式,防止在某些平台或批量脚本处理中不能正常处理。

2. 数据整理分类

在采集数据的时候会有不同场景,不同风格下的数据,这些不同来源的数据需要分开存储,不能混在一起,因为在训练的时候,不同数据集的比例会对训练模型的结果产生很大影响。对于同一个任务却不同来源的数据,比如室内、室外采集的人像数据,最好分文件夹存放。数据集包括训练集和测试集,平时使用时数据集、训练集、测试集需要以 3 个文件夹分别存储,方便进行个性化的打包与传播。

3. 归一化

归一化可用于保证所有维度上的数据都在一个变化幅度上。通常我们可以使用两种方法来实现归一化:一种是最值归一化,比如将最大值归一化成 1,最小值归一化成 -1;或者将最大值归一化成 1,最小值归一化成 0。另一种是均值方差归一化,一般是将均值归一化成 0,方差归一化成 1。

4. 数据去噪

采集数据的时候通常无法严格控制来源,比如我们常用爬虫来抓取数据,可能采集到的数据会存在很多噪声。例如,用搜索引擎采集猫的图片,采集到的数据可能会存在非猫的图片,这时候就需要人工或者使用相关的检测算法去除不符合要求的图片。数据的去噪一般对数据的标注工作会有很大帮助,能提高标注的效率。

5. 数据去重

采集到重复的数据是经常遇到的问题,比如在各大搜索引擎抓取同一类图片就会有重复数据,还有依靠视频切分成图片来获取图片的方法,数据重复性会更严重。大量的重复数据会对训练结果产生影响甚至造成模型过拟合,因此需要依据不同的任务采用不同的数据去重方案。对于图像任务来说,最简单的有逐像素比较去掉完全相同的图片,或者利用各种图像相似度算法去除相似图片。

4.3.4 数据增强

数据增强的意思是通过对训练集进行各种变换方法得到新的训练数据集,这个新的训练集具有更多数量、更多特征和更多干扰,用数据增强过的训练集可以得到泛化能力更强的模型。通常在我们训练数据集数量较小时,模型对于数据更容易发生过拟合现象,这时采用数据增强就能在一定程度上减轻网络的过拟合现象。数据增强可以分为有监督的数据增强和无监督的数据增强方法,其中有监督的数据增强又分为单样本数据增强和多样本数据增强,无监督的数据增强分为生成新的数据和学习增强策略两个方向。

1. 有监督的数据增强

(1) 几何变换

包括旋转、翻转、裁剪、变形、缩放等各类操作。值得注意的是,在一些竞赛中进行模型测试时,一般都是裁剪输入的多个版本然后将结果进行融合,对预测的改进效果非常明显。

（2）颜色变换

上面的几何变换,没有改变图像本身的内容,它可能是选择了图像的一部分或者像素进行了重分布,如果要改变图像本身的内容,那就属于颜色变换类的数据增强了,常见的包括噪声、模糊、颜色变换、擦除、填充等。基于噪声的数据增强就是在原来图片的基础上,随机叠加一些噪声,最常见的就是高斯噪声,颜色变换的另一个重要变换是颜色扰动,就是在某一个颜色空间通过增加或减少某些颜色分量,或者改变颜色通道的顺序等。

（3）SMOTE

SMOTE通过人工合成新样本来处理样本不平衡问题,从而提升分类器性能。SMOTE方法是基于插值的方法,它可以为小样本类合成新的样本。

（4）SamplePairing

SamplePairing方法的原理非常简单,从训练集中随机抽取两张图片分别经过几何或者颜色变换操作处理后经像素以取平均值的形式叠加合成一个新的样本,标签为原样本标签中的一种。这两张图片甚至不限制为同一类别,这种方法对于医学图像比较有效。

（5）Mixup

随机抽取两个样本进行简单的随机加权求和,同样样本标签也应加权求和。

（6）Cotout

随机地将样本中的部分区域去掉,并且填充0像素值,分类的结果不变。其出发点和随机擦除一样,也是模拟遮挡,目的是提高泛化能力。操作时,随机选择一个固定的正方形区域,然后采用全0填充,当然为了避免填充0值对训练的影响,应该要对数据进行中心归一化操作,归一化到0。cutout是一种类似于dropout的正则化方法,对提高模型准确率十分有效。

（7）CutMix

将一部分区域剪掉但不填充0像素而是随机填充训练集中的其他数据的区域像素值,分类结果按一定的比例分配CutMix的操作使得模型能够从一幅图像上的局部视图上识别出两个目标,提高训练的效率。

（8）Mosaic

Mosaic数据增强利用了四张图片,对四张图片进行拼接,每一张图片都有其对应的框,将四张图片拼接之后就获得一张新的图片,同时也获得这张图片对应的检测框,然后检测框,将这样一张新的图片传入神经网络中去学习,相当于一下子传入四张图片进行学习。

2. 无监督的数据增强

无监督的数据增强方法包括两类：

①通过生成模型学习数据的分布,随机生成与训练集分布一致的图片,代表方法为生成对抗网络。

②通过模型学习出适合当前任务的数据增强方法,代表方法为AutoAugment等。

单元任务13　实现简单的图像数据增强

步骤1：准备图片数据。

从数据资源服务器下载一张图片作为待转换的原始数据,如图4.16所示。

```
wget http://172.16.33.72/dataset/demo/u4/u4_Pomeranian.jpg
```

图 4.16 原始图片

步骤 2:读取图片数据。

本任务选用 tf2.1 环境进行实验操作,新建 data_aug.py 程序文件,利用 TensorFlow 读取图片数据。

```
import os
import tensorflow as tf
import matplotlib.pyplot as plt

os.environ['CUDA_VISIBLE_DEVICES'] = '0'
os.environ['TF_FORCE_GPU_ALLOW_GROWTH'] = 'true'
os.environ['TF_CPP_MIN_LOG_LEVEL'] = '3'
IMAGE_PATH = 'u4_Pomeranian.jpg'

def load_image(path):
    #读取图片
    image = tf.io.read_file(path)
    image = tf.image.decode_jpeg(image, channels=3)
    return image

if __name__ == '__main__':
    image = load_image(IMAGE_PATH)
    #输出解码之后的数据形状
    print("(height,width,channel) = ", image.shape)
```

输出图片原始尺寸:

```
(height,width,channel) = (452,500,3)
```

步骤 3:水平翻转图片。

将图片进行水平翻转,水平翻转图片如图 4.17 所示。

```
def img_flip(img):
    img = tf.image.flip_left_right(img)
    plt.figure()
    plt.imshow(img)
    plt.axis('off')
    plt.savefig('img_flip.jpg',bbox_inches = 'tight',pad_inches =0)
    print("完成！水平翻转图片保存为 img_flip.jpg")

if __name__ == '__main__':
    image = load_image(IMAGE_PATH)
    img_flip(image)
```

图 4.17　水平翻转图片

步骤4：调整图片尺寸。

将图片重新调整尺寸，大小为 200×500，如图 4.18 所示。

```
def img_resize(img):
    img = tf.image.resize(img,size = (200,500))
    img = tf.cast(img,tf.int32)
    plt.figure()
    plt.imshow(img)
    plt.axis('off')
    plt.savefig('img_resize.jpg',bbox_inches = 'tight',pad_inches =0)
    print("完成！调整尺寸图片保存为 img_resize.jpg")

if __name__ == '__main__':
    image = load_image(IMAGE_PATH)
    img_resize(image)
```

步骤5：裁剪图片。

将图片从 200×300 处裁剪为新图片，如图 4.19 所示。

图 4.18 调整图片尺寸

```
def img_crop(img):
    img = tf.image.resize_with_crop_or_pad(img,200,300)
    img = tf.cast(img,tf.int32)
    plt.figure()
    plt.imshow(img)
    plt.axis('off')
    plt.savefig('img_crop.jpg',bbox_inches = 'tight',pad_inches =0)
    print("完成！裁剪图片保存为 img_crop.jpg")

if __name__ == '__main__':
    image = load_image(IMAGE_PATH)
    img_crop(image)
```

图 4.19 裁剪图片

步骤 6：填充图片。

将图片高和宽以 0 填充为 500 和 800 的尺寸，如图 4.20 所示。

```
def img_pad(img):
    img = tf.image.resize_with_crop_or_pad(img,500,800)
    img = tf.cast(img,tf.int32)
    plt.figure()
    plt.imshow(img)
    plt.axis('off')
    plt.savefig('img_pad.jpg',bbox_inches = 'tight',pad_inches =0)
```

```
        print("完成! 填充图片保存为 img_pad.jpg")

if __name__ == '__main__':
    image = load_image(IMAGE_PATH)
    img_pad(image)
```

图 4.20　填充图片

步骤 7:调整图片亮度。

随机改变图片的亮度,如图 4.21 所示。

```
def img_brightness(img):
    img = tf.image.random_brightness(img,0.7)
    img = tf.cast(img,tf.int32)
    plt.figure()
    plt.imshow(img)
    plt.axis('off')
    plt.savefig('img_brightness.jpg', bbox_inches = 'tight',pad_inches = 0)
    print("完成! 调整亮度图片保存为 img_brightness.jpg")

if __name__ == '__main__':
    image = load_image(IMAGE_PATH)
    img_brightness(image)
```

图 4.21　调整图片亮度

步骤8：调整图片色调。

随机改变图片的色调，如图4.22所示。

```
def img_hue(img):
    img = tf.image.random_hue(img,0.5)
    img = tf.cast(img,tf.int32)
    plt.figure()
    plt.imshow(img)
    plt.axis('off')
    plt.savefig('img_hue.jpg',bbox_inches = 'tight',pad_inches = 0)
    print("完成！调整色调图片保存为 img_hue.jpg")

if __name__ == '__main__':
    image = load_image(IMAGE_PATH)
    img_hue(image)
```

图 4.22　调整图片色调

步骤9：调整图片饱和度。

随机改变图片的饱和度，如图4.23所示。

```
def img_saturation(img):
    img = tf.image.random_saturation(img,10,15)
    img = tf.cast(img,tf.int32)
    plt.figure()
    plt.imshow(img)
    plt.axis('off')
    plt.savefig('img_saturation.jpg', bbox_inches = 'tight',pad_inches = 0)
    print("完成！调整饱和度图片保存为 img_saturation.jpg")

if __name__ == '__main__':
    image = load_image(IMAGE_PATH)
    img_saturation(image)
```

图 4.23 调整图片饱和度

小　结

本单元系统地介绍了深度学习中的数据集,从数据与深度学习的关系、几大重要方向的数据集、数据的增强方法及数据标注和整理等方面进行详细讲解。本单元要求读者了解深度学习常用数据集及数据预处理方法,熟练掌握常用数据增强方法。

练　习

对一张图像进行简单的图像增强操作,包括水平翻转、调整图像尺寸、裁剪、填充图片和调整图片亮度、色调、饱和度的增强方法。

单元5 图像分类

在计算机视觉领域,图像的分类识别,可以说是最基础,最常见的一个问题,从之前的手动特征提取结合传统的分类模型,到如今的深度学习,各个图像分类库的识别率在不断被刷新。从常见物体识别,到细粒度物体识别,再到人脸识别,各个细分的图像识别领域都在取得不断进步。目前,基于深度学习下的图像分类任务,计算机的识别准确度已经超越了人眼的极限。

本单元系统地介绍了图像分类任务的基础知识,包括图像分类问题、图像分类评测指标与优化目标以及图像分类的挑战,希望学生通过本单元的学习能够掌握图像分类的基础知识,并能够熟练使用深度学习框架解决实际场景的图像分类问题。本单元的知识导图如图5.1所示。

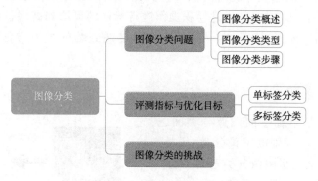

图 5.1 知识导图

课程安排

课程任务	课程目标	安排课时
熟悉图像分类问题	了解图像分类定义、图像分类问题类型以及常见图像分类方法步骤	1
熟悉图像分类的评测指标与优化目标	从单标签分类和多标签分类两个维度掌握图像分类的评价方法以及优化目标	1
了解图像分类的挑战	了解图像分类方法的瓶颈及挑战	2

5.1 图像分类问题

图像分类,顾名思义,是一个输入图像,输出对该图像内容分类的描述问题。它是计算机视觉的核心,实际应用广泛。图像分类的传统方法是特征描述及检测,这类传统方法可能对于一些简单的图像分类是有效的,但由于实际情况非常复杂,传统的分类方法不堪重负。现在,我们不再试图用代码来描述每一个图像类别,决定转而使用机器学习的方法处理图像分类问题。主要任务是给定一个输入图片,将其指派到一个已知的混合类别中的某一个标签。

5.1.1 图像分类概述

图像相比文字能够提供更加生动、容易理解及更具艺术感的信息,图像分类是根据图像的语义信息将不同类别图像区分开来,是图像检测、图像分割、物体跟踪、行为分析等其他高层视觉任务的基础。图像分类在安防、交通、互联网、医学等领域有着广泛的应用。

一般来说,图像分类通过手工提取特征或特征学习方法对整个图像进行全部描述,然后使用分类器判物体类别,因此如何提取图像的特征至关重要。基于深度学习的图像分类方法,可以通过有监督或无监督的方式学习层次化的特征描述,从而取代了手工设计或选择图像特征的工作。

深度学习模型中的卷积神经网络(CNN)直接利用图像像素信息作为输入,最大限度上保留了输入图像的所有信息,通过卷积操作进行特征的提取和高层抽象,模型输出直接是图像识别的结果。这种基于"输入-输出"直接端到端的学习方法取得了非常好的效果。

5.1.2 图像分类类型

总体来说,对于单标签的图像分类问题,它可以分为跨物种语义级别的图像分类,子类细粒度图像分类,以及实例级图像分类三大类别。

所谓跨物种语义级别的图像分类,它是在不同物种的层次上识别不同类别的对象,比较常见的包括如猫狗分类等。这样的图像分类,各个类别之间因为属于不同的物种或大类,往往具有较大的类间方差,而类内则具有较小的类内误差。图5.2是跨物种语义级别的图像分类示意图。

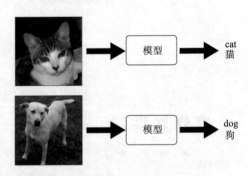

图5.2 跨物种语义级别的图像分类

CIFAR-10 的分类任务,可以看作一个跨物种语义级别的图像分类问题,如图 5.3 所示。类间方差大,类内方差小。CIFAR 包含 10 个类别,分别是 airplane、automobile、bird、cat、deer、dog、frog、horse、ship、truck,其中 airplane、automobile、ship、truck 都是交通工具,bird、cat、deer、dog、frog、horse 都是动物,可以认为是两个大的品类。而交通工具内部、动物内部,都是完全不同的物种,这些都是语义上完全可以区分的对象。

图 5.3 CIFAR-10 的分类任务

子类细粒度图像分类,相对于跨物种的图像分类,级别更低一些。它往往是同一个大类中的子类的分类,如不同鸟类的分类,不同狗类的分类,不同车型的分类等。图 5.4 所示为细粒度图像分类——花卉识别的效果,要求模型可以正确识别花的类别。

图 5.4 子类细粒度图像分类

5.1.3 图像分类步骤

1. 输入

输入是一个由 N 张图片组成的集合,每张图片都给了一个特定的类别标签,这样的数据为训练集。

2. 学习

目的是让模型学习这些图片大概是什么样子,然后记下来。称这一步为训练模型。

3. 评估

目的是检验训练模型是否有效。方法是利用此模型预测一组新数据,将预测结果与标准结果进行对比,判断模型好坏。

图像分类的整体步骤如图 5.5 所示。

图 5.5 图像分类步骤

5.2 评测指标与优化目标

5.2.1 单标签分类

在单标签图像分类任务中 Softmax loss 被广泛使用。Softmax loss 是由 softmax 和交叉熵(cross-entropy loss)loss 组合而成,全称是 softmax with cross-entropy loss,在 TensorFlow 开源框架的实现中,直接将两者放在一个层中,可以让数值计算更加稳定,因为正指数概率可能会有非常大的值。softmax loss 的定义如下:

$$L = -\sum_{j=1}^{T} y_j \log S_j$$

损失函数的定义如上,其是 softmax 函数的输出向量的第 j 个值,表示这个样本输入这个类别的概率,y_j 的取值为 0 或者 1,当得到分类结果正确(最大概率值对应的标签 = 真实标签)时,取值为 1,其他情况取值为 0。当正确分类项得到的概率值越低时,对应的损失函数也就越大。

原始的 Softmax loss 非常优雅、简洁,被广泛用于分类问题。它的优点是优化类间的距离,但优化类内距离时比较弱。对于特定问题,可以对原始函数进行改进,产生多类变种损失函数。

在单标签比赛中常用的准确度评测指标是 top-N 准确率,其中 ImageNet 挑战赛中常用的是 top-5 和 top-1 准确率。所谓 top-1,即统计样本概率最大的类是否为对应的类别,如是则代表分类正确,反之代表分类错误。top-5 则代表分类的前 5 个类别中是否包含了正确的类别,如果正确则代表分类正确。由此可知,top-5 的指标必定高于或等于 top-1。

5.2.2 多标签分类

多标签分类就是给每个样本一系列的目标标签,可以想象成一个数据点的各属性不是相互排斥的(一个水果既是苹果又是梨就是相互排斥的),比如一个文档相关的话题,一个文本可能被同时认为是农业、金融或者教育相关话题。多标签分类方法的本质是把一个多标签问题转化为每个标签上的二分类问题。常用的策略有两种:一对一策略和一对多策略。

一对一策略:给定数据集 D 中有 N 个类别,这种情况下就是将这些类别两两配对,从而产生 $N(N-1)$ 个二分类任务,在测试时把样本交给这些分类器,然后进行投票。

一对多策略:将每一次的一个类作为正例,其余作为反例,总共训练 N 个分类器。测试时若仅有一个分类器预测为正的类别则对应的类别标记作为最终分类结果,若有多个分类器预测为正类,则选择置信度最大的类别作为最终分类结果。

5.3　图像分类的挑战

影响图像分类识别的因素很多,最常见的就是光照、形变、尺度、遮挡、模糊等,这些可统称为一般因素。

还有一大类因素就是同一类别的东西,形态各异,比如椅子类别,所有的图片都可以被认为是椅子,但是要让分类模型去识别这些很难,这类问题可以归结为类内差异太大。

较常见的还有细粒度分类,就是可能都属于一个大类,但是细分下来,就存在细微的差异,细粒度分类也是图像分类识别的一个分支,比较常见的分类就是对狗、鸟,以及飞机的各种型号的分类,这类问题就是类间差异太小。总的来说,影响分类识别的因素可以归结为三个方面:

①比较常见的基于图像本身的一些因素,如光照、形变、尺度、遮挡、模糊等,如图5.6所示。

图 5.6　基于图像本身的因素

②类内差异太大,如椅子、桌子,虽然都叫椅子、桌子,可是形态各异,如图5.7所示。

图 5.7　类内差异大的因素

③类间差异太小,最常见的就是细粒度分类。

图 5.8 细粒度分类

第一种因素,所有分类识别问题都经常遇到,这是因为真实环境就是各种各样的;第二种因素,很多时候都是因为类别定义的问题,人们对物体分类并不是基于其形状或者样子,而是基于功能,这种基于功能的抽象的类别定义,让分类模型对分类对象的理解更加困难;第三种因素,是属于细粒度分类的问题。

考虑最常见的分类情况,如对常见交通工具以及各种常见动物的分类等,真实的自然场景总是复杂多变的,光线有强有弱,形状也会有各种变化,因为换个视角观察,同一个物体呈现在图像上的形状会发生变化,尺度有大有小,因为拍摄的远近距离不同,同一个物体呈现在图像上的大小也会不同,更不用说模糊、遮挡等因素,所以真实的场景就是包含了各种可能。

分类识别模型的训练及预测,一般通用流程就是给定一个训练集,在这个训练集上训练了一个模型,然后将这个模型在测试集上进行预测,计算出一个准确率。然而,很多模型,换一个数据集,哪怕还是同样的类别,性能立马就会大打折扣,这个对于人类来说,会觉得不可思议,还是同样的狗,为什么换几张图,模型就会失效,分类结果出现其他匪夷所思的东西。所以说,模型在某些方面的"聪明"和"愚蠢"经常是同时存在的。模型的不稳定,或者不够鲁棒,是目前分类中最常遇到的问题,我们经常会看到对某张图稍微修改几个像素值,这张图在模型眼里就已经千差万别了。人类的视觉感知,相对来说会稳定得多。如果仔细分析,就会发现,分类本质上还是一个特征映射的过程,从高维的图像空间映射到高维的特征空间,然后再映射到标签空间,网络经过一系列复杂的映射,将同一类图像聚集到一起。

人类的视觉感知,比 CNN 这些模型要鲁棒得多,因为人类的视觉感知,不是简单的一个函数映射,而是经过了漫长的进化、演化而来的,融合了目标检测、背景过滤、联想、决策、推理等因素,交织在一起的一个非常精细的智能系统。不是一个简单的 CNN 网络就能轻易替代的,目前的分类模型,可能会在某个非常细化的领域,识别水平会表现很好,可是这些分类模型都很"单一",无法将这种认知能力迁移到其他地方。

单元任务 14　102 种花卉图像分类实战

本任务要求利用深度学习构架神经网络模型解决 102 种花卉的分类任务,数据集来自世界各地共 102 个品种的花卉。数据集可以从数据资源服务器下载。本任务使用 tf2.1 开发环境。

```
#从数据资源平台下载数据集并解压
wget http://172.16.33.72/dataset/Classification_102flowers.tar.gz
tar -zxvf Classification_102flowers.tar.gz
rm Classification_102flowers.tar.gz
```

将数据集的目录展开,可以详细看到部分样本数据,如图 5.9～图 5.11 所示,其中 cat_to_name.json 是一个花卉文件标签映射类名文件,有 102 个元素,分别是从 1～102 映射到对应的花卉品种名称。test 文件夹是测试集数据,其中是 819 张测试样本图片。train 文件夹是训练集数据,其中是从 1～102 的子目录,分别对应着 102 种花卉。test 文件夹是验证集数据,其中是从 1～102 的子目录,子目录下全是相应花卉的图片数据。

```
dataset/
├──cat_to_name.json
├──test
├──train
└──valid
```

图 5.9 测试集图片(一)

图 5.10 测试集图片(二)

图 5.11 测试集图片(三)

步骤 1:数据读取和预处理。

创建数据读取文件 dataset.py,将在该程序中完成数据集的读取、数据标签的编码,以及载入原始数据为数据管道。

```python
import tensorflow as tf
import glob
import json
import random

TRAIN_IMAGE_PATH = "dataset/train/*/*.jpg"
VAL_IMAGE_PATH = "dataset/valid/*/*.jpg"
TEST_IMAGE_PATH = "dataset/test/*.jpg"
LABEL_JSON_PATH = "dataset/cat_to_name.json"
IMAGE_RESIZE = (512,512)
AUTOTUNE = tf.data.experimental.AUTOTUNE

def load_label():
    #读取cat_to_name.json获取花卉品类名
    with open(LABEL_JSON_PATH,'r') as f:
        label_json = json.load(f)
        index_to_name = dict((int(index)-1,name) for index,name in label_json.items())
        name_to_index = dict((name,index) for index,name in index_to_name.items())
    return (index_to_name,name_to_index)

def load_image(path,label):
    #处理一张图片
    image = tf.io.read_file(path)
    image = tf.image.decode_jpeg(image,channels=3)
    image = tf.image.resize(image,IMAGE_RESIZE)
    image = tf.cast(image,tf.float32) / 255
    return image,label

def load_dataset(batch_size):
    #载入数据为数据管道
    train_image_path = glob.glob(TRAIN_IMAGE_PATH)
    val_image_path = glob.glob(VAL_IMAGE_PATH)
    #打乱数据
    random.shuffle(train_image_path)
    random.shuffle(val_image_path)
    #print(train_image_path[:3])

    #标签编码
    train_image_label = [int(a.split('/')[2])-1 for a in train_image_path]
    val_image_label = [int(b.split('/')[2])-1 for b in val_image_path]
    #print(train_image_label[:3])
    #训练集数据管道
    train_ds = tf.data.Dataset.from_tensor_slices((train_image_path,train_image_label))
    train_ds = train_ds.map(load_image,num_parallel_calls=AUTOTUNE)
    train_ds = train_ds.shuffle(buffer_size=2000).repeat().batch(batch_size)
    train_ds = train_ds.prefetch(buffer_size=AUTOTUNE)
```

```python
    #验证集数据管道
    val_ds = tf.data.Dataset.from_tensor_slices((val_image_path,val_image_label))
    val_ds = val_ds.map(load_image,num_parallel_calls = AUTOTUNE)
    val_ds = val_ds.batch(batch_size)
    val_ds = val_ds.prefetch(buffer_size = AUTOTUNE)

    return train_ds,val_ds

def load_step(batch_size):
    #加载训练集和验证集步数
    train_image_path = glob.glob(TRAIN_IMAGE_PATH)
    val_image_path = glob.glob(VAL_IMAGE_PATH)
    steps_per_epoch = len(train_image_path) //batch_size
    validation_steps = len(val_image_path) //batch_size
    return steps_per_epoch,validation_steps
```

步骤2：模型构建。

创建模型文件model.py，编写一个以VGG16为主干网络的模型。

```python
import tensorflow as tf


def build_model(backbone_freeze = False):
    #构建模型
    backbone = tf.keras.applications.VGG16(weights = 'imagenet',include_top = False, input_shape = (512,512,3))
    backbone.trainable = backbone_freeze
    model = tf.keras.Sequential()
    model.add(backbone)
    model.add(tf.keras.layers.GlobalAveragePooling2D())
    model.add(tf.keras.layers.Dense(512,activation = 'relu'))
    model.add(tf.keras.layers.Dropout(0.5))
    model.add(tf.keras.layers.Dense(256,activation = 'relu'))
    model.add(tf.keras.layers.Dropout(0.5))
    model.add(tf.keras.layers.Dense(102))
    return model
```

步骤3：模型训练。

创建训练程序train.py，这里使用最简单的fit方法进行训练，模型保存在saved_model/1/目录下，logs目录下记录TensorBoard训练日志文件，程序运行结果如图5.12和图5.13所示。

```python
import os
import tensorflow as tf
import datetime
from dataset import load_dataset,load_step
from model import build_model


os.environ['CUDA_VISIBLE_DEVICES'] = '0'
os.environ['TF_FORCE_GPU_ALLOW_GROWTH'] = 'true'
```

```python
os.environ['TF_CPP_MIN_LOG_LEVEL'] = '3'
BATCH_SIZE = 16
EPOCH = 20
MODEL_PATH = "./data/model.h5"

print("加载数据>>>")
train_ds, val_ds = load_dataset(BATCH_SIZE)
steps_per_epoch, validation_steps = load_step(BATCH_SIZE)
#print(train_ds)
print("加载模型>>>")
model = build_model()
model.summary()
#定义 TensorBoard 回调
log_dir = os.path.join(
    'logs', datetime.datetime.now().strftime("%Y%m%d-%H%M%S")) tensorboard_callback = tf.keras.callbacks.TensorBoard(
    log_dir = log_dir, histogram_freq = 1)


def train():
    tf.print("开始训练>>>")
    #定义优化器
    optimizer = tf.keras.optimizers.Adam(0.0003)
    #定义损失
    loss = tf.keras.losses.SparseCategoricalCrossentropy(from_logits = True)
    #编译模型
    model.compile(optimizer = optimizer, loss = loss, metrics = ['accuracy'])
    #训练模型
    model.fit(train_ds, steps_per_epoch = steps_per_epoch, epochs = EPOCH,
        validation_data = val_ds, validation_steps = validation_steps,
        callbacks = [tensorboard_callback])
    tf.print("训练完成!")
    model.save(MODEL_PATH, save_format = 'h5')

train()
```

图 5.12　训练模型(一)

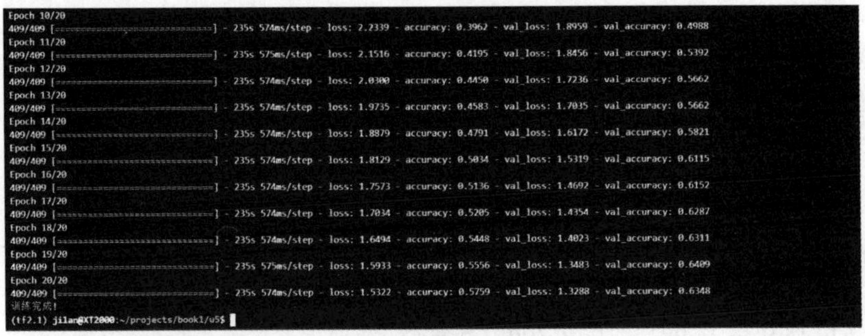

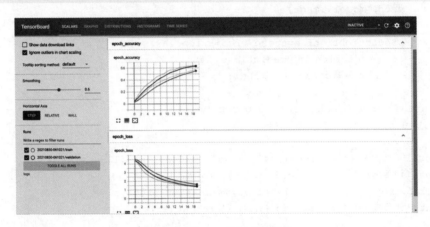

图 5.13 训练模型(二)

使用 TensorBoard 可视化训练日志,在命令行输入启动命令,使用浏览器访问 TensorBoard 服务地址和端口,如图 5.14 所示。

```
tensorboard --host 172.16.33.106 --port 8888 --logdir logs
```

图 5.14 TensorBoard 可视化

从训练集和验证集的损失、准确率的曲线趋势可以看出,模型仍然处于欠拟合状态,读者可以自行增大 epoch 值继续训练。

步骤 4:模型评估。

创建模型评估程序 eval.py 文件,在模型评估中,需要通过模型对验证集的预测结果和真实结果进行 Top-1-error 和 Top-5-error 两项指标的评估。Top-1-error 指标意思是在模型的预测结果中,输出第一个预测结果的正确率,而 Top-5-error 则是在模型的预测结果中,输出前五项的预测结果的正确率,程序运行结果如图 5.15 所示。

```
import os
import tensorflow as tf
from dataset import load_dataset,load_step
from tqdm import tqdm

os.environ['CUDA_VISIBLE_DEVICES'] = '0'
os.environ['TF_FORCE_GPU_ALLOW_GROWTH'] = 'true'
os.environ['TF_CPP_MIN_LOG_LEVEL'] = '3'
```

```python
BATCH_SIZE = 1
MODEL_PATH = "./data/model.h5"

#加载数据
print("加载数据>>>")
_,val_ds = load_dataset(BATCH_SIZE)
_,validation_steps = load_step(BATCH_SIZE)
#重载模型
print("重载模型>>>")
model = tf.keras.models.load_model(MODEL_PATH)

def top_k_error(model,val_ds,batch_num,k):
    #top_k_error评估函数
    count = 0
    tf.print("开始评估>>>")
    for image,label in tqdm(val_ds,total = batch_num):
        pred = model(image)
        predictions = (tf.math.top_k(pred,k = k).indices.numpy())
        predictions = tf.squeeze(predictions,axis = 0)
        for prediction in predictions:
            if prediction = = label:
                count + = 1
    error = (batch_num - count) / batch_num
    return error

error_1 = top_k_error(model,val_ds,validation_steps,1)
print('评估完成! Top-1-error: {:.2%}'.format(error_1))
error_5 = top_k_error(model,val_ds,validation_steps,5)
print('评估完成! Top-5-error: {:.2%}'.format(error_5))
```

图 5.15 使用测试集评估模型

从评估结果可以得知,该模型在 Top-1-error 上的错误率为 36.55%,在 Top-5-error 上的错误率下降为 10.88%,可知模型对于 102 种花卉具有良好的预测性能。

步骤 5:模型测试。

创建一个分类器 classify.py,使用保存的模型对未知种类的花卉进行预测,返回预测结果,如图 5.16 所示。

```python
import os
import tensorflow as tf
import numpy as np
import matplotlib.pyplot as plt
```

```python
from dataset import load_label

os.environ['CUDA_VISIBLE_DEVICES'] = '0'
os.environ['TF_FORCE_GPU_ALLOW_GROWTH'] = 'true'
os.environ['TF_CPP_MIN_LOG_LEVEL'] = '3'
IMAGE_PATH = "dataset/test/image_00132.jpg"
IMAGE_RESIZE = (512,512)
MODEL_PATH = "./data/model.h5"

def load_image(path):
    #处理一张图片
    image = tf.io.read_file(path)
    image = tf.image.decode_jpeg(image,channels=3)
    image = tf.image.resize(image,IMAGE_RESIZE)
    image = tf.cast(image,tf.float32) / 255
    image = tf.expand_dims(image,axis=0)
    return image

#加载数据
print("加载数据>>>")
image = load_image(IMAGE_PATH)
index_to_name,name_to_index = load_label()
#重载模型
print("重载模型>>>")
model = tf.keras.models.load_model(MODEL_PATH)

result = model.predict(image)
prediction = np.argmax(result)
prediction = index_to_name.get(prediction)
print("测试完成!")

img = tf.io.read_file(IMAGE_PATH)
img = tf.image.decode_jpeg(img,channels=3)
plt.figure()
plt.title('Prediction: % s' % prediction)
plt.imshow(img)
plt.axis('off')
plt.savefig('data/img001.jpg',bbox_inches='tight',pad_inches=0)
```

测试完成,保存图片为 data 目录下的 img001.jpg,打开后可以看到模型预测结果为 passion flower(西番莲)。

任务总结:本任务完成了一个实际的图像分类任务,通过这次任务实践操作,熟悉了完成一个图像分类的主要步骤,通过构建模型和训练模型,获取了较为精准的对 102 种花卉的预测模型,并且完成了模型评估和模型测试。

图 5.16 模型预测结果

小　结

本单元系统地介绍了图像分类任务的相关基础知识,从图像分类问题、图像分类评测指标与优化目标以及图像分类面临的挑战等方面系统地梳理了图像分类任务的整体架构。读者可以通过本单元的学习掌握图像分类基础知识,并能够在单元任务的实践中掌握图像分类的数据处理和深度学习算法。

练　习

完成 285 种鸟的分类,数据集为数据资源平台的 Classification_285bird.tar.gz。

单元6 目标检测

目标检测是图像识别的主要任务之一,作为计算机视觉和数字图像处理的核心任务,目标检测具有长期的理论研究价值和广泛的实际应用场景。目标检测任务中包含有两个子任务:第一输出图像中感兴趣目标的类别信息;第二输出目标物体的具体位置信息。目标检测具有多种经典算法,总体发展历程可以总结为从传统图像处理算法到深度学习目标检测算法的过程。传统的图像处理算法是利用数字图像处理技术对图像进行浅层处理后提取图像特征,再借助分类器完成检测任务。深度学习目标检测算法分为两条技术路线,分别为二阶段目标检测算法和一阶段目标检测算法。二阶段目标检测算法检测精度高、检测速度慢、实时性较差,代表性的算法是 R-CNN 系列算法。一阶段目标检测算法具有良好的检测精度和优越的实时性,因而在工业应用中更为普遍,典型算法有 YOLO 系列和 SSD 系列。在实际生产生活应用中,目标检测具有极为广泛的应用场景。在自动驾驶、智能监控、工业检测、机器人导航、航空航天等科技领域均有广泛应用。

本单元从目标检测的综述开始阐述了目标检测算法的发展历程和主要典型的检测算法,然后分别从数据集、评测指标和损失函数介绍目标检测的基础,要求学生基本掌握目标检测的算法原理,最后在典型算法的实例任务中展示基本的人工智能目标检测项目开发流程,帮助读者进一步深入理解目标检测算法的具体实现。本单元的知识导图如图6.1所示。

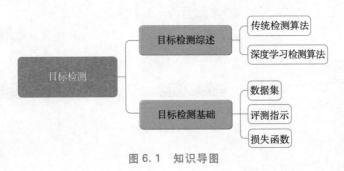

图 6.1 知识导图

课程安排

课程任务	课程目标	安排课时
目标检测综述	帮助学生深刻理解目标检测的具体内容,了解目标检测算法发展历程和具有代表性的典型算法	1

续表

课程任务	课程目标	安排课时
目标检测基础	熟悉目标检测的数据集构成、评测指标、常用分类和定位损失函数	4
Yolov3 算法	整体理解 Yolov3 算法的框架,熟悉用 TensorFlow 编写 Yolov3 模型的代码	4
SSD 算法	整体理解 SSD 算法完成目标检测任务的模型,能够使用 TensorFlow 完成 SSD 模型代码的编写	4

6.1 目标检测综述

目标检测具有多种经典算法,总体发展历程可以总结为从传统图像处理算法到深度学习目标检测算法的过程。传统的图像检测算法是利用滑动窗口对处理后的图像提取特征,再借助分类器完成检测任务,其典型算法有 V-J 检测器、HOG 特征提取算法和 DPM 特征提取算法。深度学习目标检测算法分为二阶段目标检测算法和一阶段目标检测算法。二阶段目标检测算法检测精度高、检测速度慢、实时性较差,代表性的算法是 R-CNN 系列算法。一阶段目标检测算法具有良好的检测精度和优越的实时性因而在工业应用中更为普遍,典型算法有 YOLO 系列和 SSD 系列。随着时间的推移,这些算法的不断推陈出新,衍生出各种改进增强的算法,不断地提升目标检测在实际应用中的准确度和运行速度等各方面性能,让目标检测这一计算机视觉领域呈现百家争鸣的发展局面。目标检测发展历程如图 6.2 所示。

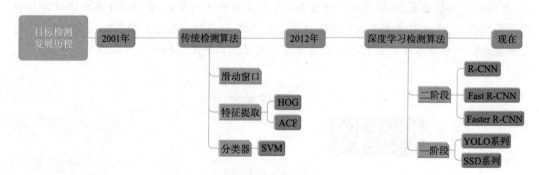

图 6.2 目标检测发展历程

6.1.1 传统检测算法

传统目标检测方法一般分三个阶段进行:首先,在给定图像上选定若干候选区域目标,最常使用的是长宽比一定的滑动窗口;然后,在这些区域中提取特征,常用的特征提取算法有 HOG、ACF 和 SIFT 等;最后,根据前面得到的特征使用训练好的分类器如支持向量机(SVM)完成分类任务。传统检测算法的局限性是:检测精度低、速度慢且整个过程分步进行。传统检测算法是以机器学习为基础、以特征工程为重心的算法,由于需要大量的手工参数设计,不仅模型设计复杂多变,而且在整体的机器学习数据拟合和分类上存在众多局限性，所以以机器学习为基础的传统算法的精确度无法有进一步的突破。之后以深度学习为基础的行人检测的出现完全取代了传统的机器学习时代手工设计的特征工程算法。图 6.3 所示为传统检测算法。

图 6.3 传统检测算法

6.1.2 深度学习检测算法

深度学习在计算机视觉领域发生重大影响的突破是，在 2012 年的 ImageNet 图像分类比赛中，Hinton 的研究组首次采用了多层卷积神经网络(CNN)、ReLU 激活函数和随机失活(Dropout)的算法来处理图像分类问题，他们分类预测结果的准确率远远大于采用传统特征工程图像处理方案的小组。随后 Ross Girshick 提出的 R-CNN 算法为目标检测领域的研究开辟了新世界。在此基础上先后有众多研究者创造出二阶段目标检测算法，如 Fast R-CNN 和 Faster R-CNN 等，奠定了人工智能时代深度学习模型目标检测的基础。

R-CNN 算法将底层的图像特征延展到高层的语义特征，用一种更为抽象的特征取代了之前以机器学习为基础的特征工程。总的模型组成为选择性搜索(Selective Search)、卷积神经网络(CNN)和支持向量机(SVM)。先用选择性搜索算法代替传统的滑动窗口，提取出 2 000 个候选区域；然后针对每个候选区域，普遍使用卷积神经网络提取特征，并通过训练可支持向量机作为分类器，将卷积神经网络提取出来的特征作为输出，得到每个候选区域属于某一类的得分；最后在每个类别上用非极大抑制(Non-maximum Suppression，NMS)来舍弃掉重复率较高的区域得到最终检测结果。R-CNN 算法虽然极具创造性地使用卷积神经网络来提取特征，但是仍然需要大量的手工设计特征的烦琐步骤。图 6.4 所示为 R-CNN 模型。

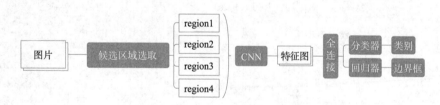

图 6.4 R-CNN 模型

R-CNN 需要非常多的候选区域以提升准确度，但其实有很多区域是彼此重叠的，因此 R-CNN 的训练和推断速度非常慢。Fast R-CNN 使用 CNN 先提取整个图像的特征，而不是从头开始对每个图像块提取多次。然后将创建候选区域的方法直接应用到提取到的特征图上。使用 RoI 池化将特征图块转换为固定大小，并馈送到全连接层进行分类和定位。因为 Fast R-CNN 不会重复提取特征，因此显著地减少了处理时间，大大提升了模型的训练速度和检测速度，让二阶段的目标检测算法逐渐成熟。图 6.5 所示为 Fast R-CNN 模型。

二阶段目标检测的范式 Faster R-CNN 将特征提取、边界候选框提取、边界锚点框回归和分类都整合在了一个综合性的神经网络中，并同样在卷积主要模块之间共享参数计算，

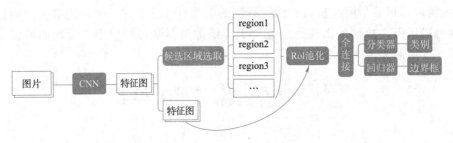

图 6.5　Fast R-CNN 模型

模型不但在整体上形成了端到端的训练和检测,而且进一步提升了模型训练和检测速度,完全具备了实际应用价值。Faster R-CNN 模型(见图 6.6)的主要组成:第一,特征提取的卷积层。特征提取的卷积层主要使用一组基础的卷积层+激活函数为 ReLU 的激活层+池化层提取图像的特征图。在特征图第一次被提取出后被共享用于后续的候选区域生成网络(Region Proposal Networks,RPN)和全连接层重复利用,避免了大量冗余的卷积计算。第二,候选区域生成网络。候选区域生成网络用于生成候选区域,如图 6.7 所示。第三,候选区域池化层。候选区域池化层收集输入的特征图数据和候选锚点框,综合长宽坐标和纵横比后提取候选特征图,输入后续全连接层判定目标类别和回归目标具体位置。

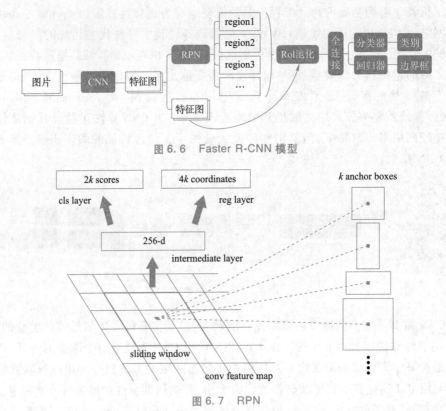

图 6.6　Faster R-CNN 模型

图 6.7　RPN

二阶段的目标检测算法形成完整的体系的同时,一阶段的检测算法也相继被提出。一阶段的目标检测算法由于抛弃了候选区域生成子网络的设计,直接从特征提取的卷积层主网络上进行分类与回归的任务,所以普遍具有更快的检测速度。一阶段的目标检测模型相比于二阶段更加注重检测的速度,从实时性和实用性出发,逐步向高精度、高速度的检测器发展,其代表性模型有 SSD、DSSD、DSOD、YOLO 系列、RetinaNet。SSD 和 YOLO 模型如图 6.8 所示。

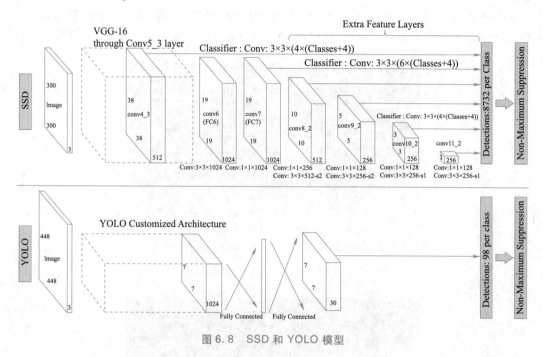

图 6.8　SSD 和 YOLO 模型

如图 6.8 所示,Yolo 系列的算法将整个特征图分割为众多网格,通过 K-Means 聚类筛选先验边界框的纵横比尺寸大小,并使用网格的参数化坐标和先验边界框与真实边界框偏移量完成目标的分类和回归任务。由于其模型简单易用且检测准确快速,故具有广泛的实际应用。

相比于二阶段的检测模型,一阶段 SSD 在速度上的提升基本来源于淘汰了候选区域边界框,以及网络后段的像素级特征重采样阶段。SSD 在精度上的提升可以归结于两点:第一,在模型预测类别和边界框位置时,卷积层普遍使用了更小尺寸的卷积滤波器;第二,模型为多尺度和不同宽高比的特征图设置了分解的独立检测器。

6.2　目标检测基础

6.2.1　数据集

目标检测的公开数据集很多,如 PASCAL VOC2007-VOC2012。PASCAL VOC 数据集来自于 PASCAL VOC 挑战赛,该挑战赛从 2005 年开始到 2012 年终止,主要有图像分类、目标检测、目标分割和行为分类等几个子任务挑战,其中以目标检测为主要挑战任务,挑战赛中官方公开的数据集就是 PASCAL VOC 数据集,最为常用的数据集有 2007 年目标检测挑战赛使用的 VOC2007 数据集和 2012 年目标检测挑战赛使用的 VOC2012 数据集。2005 年发布的 PASCAL 数据集检测类别只有 4 类,从 2007 年开始增加到 20 类,并且引入了图像分割标注和人体结构布局标注。随着官方标注信息和评估指标的完善,VOC2007 和 VOC2012 逐渐成为了公开的常用数据集,大部分图像分类,目标检测和图像分割任务都可以在这两个数据集上完成训练和评估的要求。VOC2007 和 VOC2012 都包括日常生活中常见的 20 种物体类别,如图 6.9 和图 6.10 所示。VOC2007 和 VOC2012 将训练集、测试集和标签文件分开,一共六个压缩文件。完整的数据集都可以在数据资源服务器中通过下载获取。

图 6.9　PASCAL VOC 数据集(一)

图 6.10　PASCAL VOC 数据集(二)

6.2.2　评测指标

目标检测的评测指标主要为平均精确度(Average Precision,AP)。平均精确度主要是由精确度与召回率平滑后的曲线面积计算所得。在多个检测类上进一步均值化就是均值平均精确度(Mean Average Precision,mAP)。

将样本根据其真实类别与预测类别的组合划分为"混淆矩阵"(Confusion Matrix),见表 6.1。预测结果可以判别为真正例(True Positive),又称真阳样本;假正例(False Positive),又称假阳样本;真反例(True Negative),又称真阴样本;假反例(False Negative),又称假阴样本。

表 6.1　混淆矩阵

真实情况	预测结果	
	正　例	反　例
正例	TP(真正例)	FN(假反例)
反例	FP(假正例)	TN(真反例)

根据真实边界框和预测边界框的交集和并集的比值可以将某一次预测值的图像划分为 TP、FP、FN 和 TN 四部分,如图 6.11 所示。

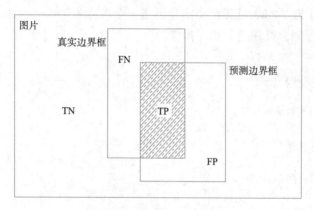

图 6.11　真实边界框和预测边界框

其中精确率 P(Precision)和召回率 R(Recall)的计算公式如下：

$$P = \frac{TP}{TP + FP}$$

$$R = \frac{TP}{TP + FN}$$

根据类别置信度阈值可以计算出多种情况下的 AP 值,$AP_{0.5}$ 表示类别置信度为 0.5 时的检测精度,也是使用比较多的标准值。

6.2.3　损失函数

目标检测需要完成分类+定位的任务,一般都包含两类损失函数,其中一个是分类损失,另一个是定位损失。这两类损失函数用在检测模型的最后端,根据模型输出的类别和边界框位置计算相应的损失值,模型的总损失则是两部分的和。

1. 类别损失

类别损失一般使用交叉熵损失(Cross Entropy Loss)。

还有改进的交叉熵损失 Focal loss,减轻了正负样本不均衡的影响,普遍使用在 RetinaNet 模型中。

2. 位置损失

- L1 loss,平均绝对误差(Mean Absolute Error,MAE)。计算为模型预测值和真实值之间距离的平均值。

- L2 loss，均方误差损失（Mean Square Error，MSE）。计算为模型预测值和真实值之差平方的平均值。
- Smooth L1 loss。改进的 L1 loss。
- IoU loss，交并比损失。预测边界框为 P，真实边界框为 G。则交并比为 IoU = $(P \cap G)/(P \cup G)$，则交并比损失为 1-IoU。

单元任务 15　使用 Yolov3 算法实现目标检测

步骤 1：创建工程和 Yolov3 虚拟环境。

使用 Anaconda 为 Yolov3 目标检测项目创建一个名称为 Yolov3、TensorFlow 版本为 2.4 的虚拟环境，打开控制台终端进行以下操作。

```
#创建工程文件
mkdir ~/projects/yolov3

#创建名为 yolov3 的虚拟环境，如图 6.12 所示
conda create-n yolov3 python=3.7
#输入 y 确认下载基础环境包

#进入 yolov3 虚拟环境，如图 6.13 所示
conda activate yolov3
```

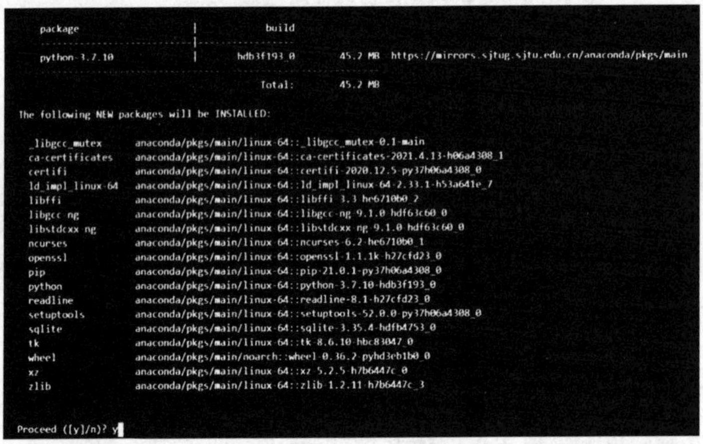

图 6.12　创建虚拟环境

图 6.13　激活虚拟环境

安装 GPU 版 TensorFlow 和其 CUDA、cudnn 依赖，在命令行输入如下代码：

```
conda install cudatoolkit=10.1 cudnn=7.6.5
#输入 y 确认下载依赖包，如图 6.14 所示和图 6.15 所示
```

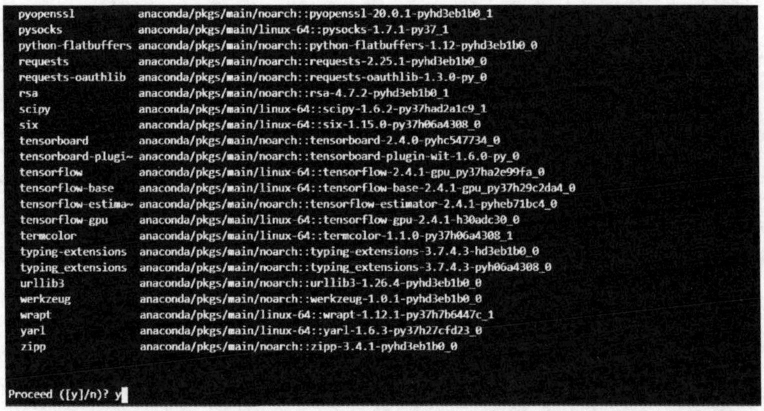

图 6.14 安装环境依赖(一)

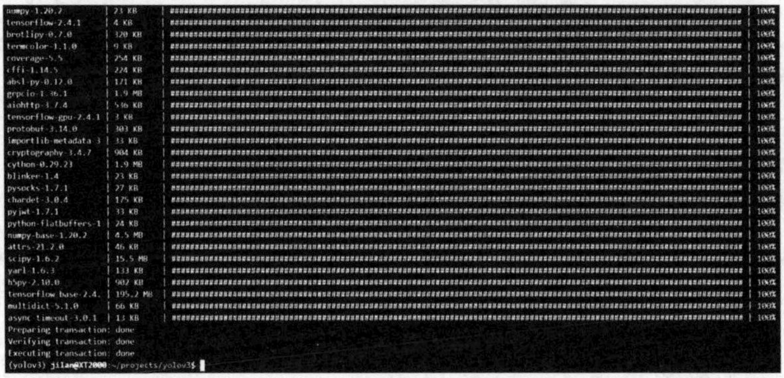

图 6.15 安装环境依赖(二)

保持控制台终端 Yolov3 环境始终处于激活状态,在工程文件夹下新建 requirements.txt 文件,写入下列软件包,保存之后在控制台终端输入 pip install-r requirements.txt,安装好之后就完成了 Yolov3 基本开发环境的软件包配置。在代码编辑器中输入以下代码:

```
matplotlib
opencv-python
tensorflow-gpu = =2.4.0
lxml
tqdm
```

运行结果如图 6.16 所示。

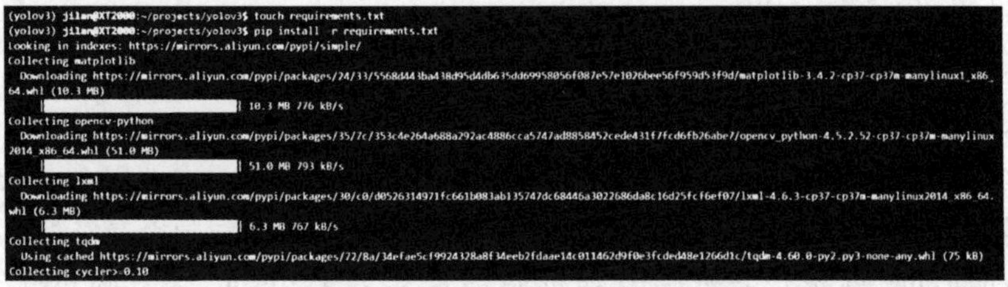

图 6.16 安装第三方包依赖

步骤 2:准备数据集。

在 Yolov3 工程文件夹下新建 datasets 数据集文件夹,首先从数据资源下载 VOC2007 和 VOC2012,在命令行输入以下代码:

```
#从数据资源平台下载所需数据集压缩包
wget zyadmin@ 172.16.33.72:/home/zyadmin/dataset/VOC/VOCtrainval_06-Nov-2007.tar
wget zyadmin@ 172.16.33.72:/home/zyadmin/dataset/VOC/VOCtest_06-Nov-2007.tar
```

下载完成后在 datasets 目录下多出三份打包文件,分别是 VOC2007 训练集和验证集:VOCtrainval_06-Nov-2007.tar,VOC2007 测试集:VOCtest_06-Nov-2007.tar。

对 VOC2007 和 VOC2012 数据集包原始文件解包,在命令行输入以下代码:

```
#继续在./datasets 文件夹下进行解包操作
#VOC2007 训练集验证集
tar -xvf VOCtrainval_06-Nov-2007.tar
#VOC2007 测试集
tar -xvf VOCtest_06-Nov-2007.tar
#VOC2012 训练集验证集
tar -xvf VOCtrainval_11-May-2012.tar
```

解包完成后在 datasets 文件夹下可以得到 VOCdevkit 文件夹,该目录下有 VOC2007 和 VOC2012 的数据集目录。解包后完整的 VOC 数据集目录如下。其中 Annotations 目录、SegmentationClass 目录和 SegmentationObject 目录为标注文件,用于目标检测和图像分割等任务。ImageSets 目录为数据集标签索引目录。JPEGImages 为图像数据集目录,该目录下全部是格式为.jpg 的图片数据。在本任务中主要用到的是 JPEGImages、ImageSets 目录和 Annotations 目录的数据。

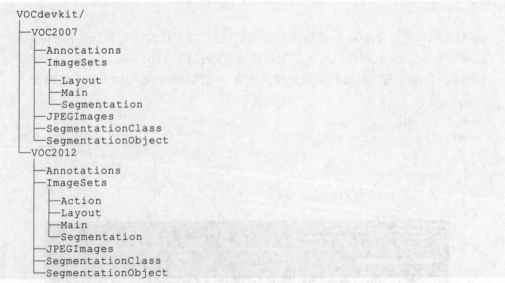

数据集载入有多种方式,这里推荐使用 TFRecord 二进制格式文件向 TensorFlow 模型载入训练数据。TFRecord 文件基于 Google Protocol Buffers 跨平台跨语言序列化结构的协议标准,TensorFlow 内核很多数据处理机制都是基于 TFRecord 文件做的优化。因此 TFRecord 文件相比于其他数据载入方式具有读取速度快、效率高的优势,其文件扩展名为.tfrecord 文件。下面分别将 VOC2007 和 VOC2012 的数据集转换成 TFRecord 格式文件。

在工程文件夹下新建 voc2007_tfrecord.py 文件,写入转换数据集程序。

首先,在工程目录下新建 data/ 目录,在命令行中输入以下代码。

```
mkdir data
```

在 ./data 目录下新建 voc2007.names 目标检测类别文件,在命令行中输入以下代码。

```
cd data/
touch voc2007.names
```

在 voc2007.names 中逐行写入目标检测的类别信息,在代码编辑器中输入以下代码。

```
aeroplane
bicycle
bird
boat
bottle
bus
car
cat
chair
cow
diningtable
dog
horse
motorbike
person
pottedplant
sheep
sofa
train
tvmonitor
```

在工程目录下新建 voc2007_tfrecord.py,voc2007_tfrecord.py 程序将 VOC2007 数据集转换为 TFRecord 二进制格式,在代码编辑器中输入以下代码。

```
import time
import os
import hashlib

from absl import app,flags,logging
from absl.flags import FLAGS
import tensorflow as tf
import lxml.etree
import tqdm

flags.DEFINE_string('data_dir','./datasets/VOCdevkit/VOC2007/',' PASCAL
                    VOC2007 dir')
flags.DEFINE_enum('split','train',['train','val'],'train or val')
flags.DEFINE_string('output_file','./data/voc2007_train.tfrecord','outpot
tfrecord dataset')
```

```python
        flags.DEFINE_string('classes','./data/voc2007.names','voc.names file')


    def build_example(annotation,class_map):
        img_path = os.path.join(
            FLAGS.data_dir,'JPEGImages',annotation['filename'])
        img_raw = open(img_path,'rb').read()
        key = hashlib.sha256(img_raw).hexdigest()

        width = int(annotation['size']['width'])
        height = int(annotation['size']['height'])
xmin = []
    ymin = []
    xmax = []
    ymax = []
    classes = []
    classes_text = []
    truncated = []
    views = []
    difficult_obj = []
    if 'object' in annotation:
        for obj in annotation['object']:
            difficult = bool(int(obj['difficult']))
            difficult_obj.append(int(difficult))

            xmin.append(float(obj['bndbox']['xmin']) / width)
            ymin.append(float(obj['bndbox']['ymin']) / height)
            xmax.append(float(obj['bndbox']['xmax']) / width)
            ymax.append(float(obj['bndbox']['ymax']) / height)
            classes_text.append(obj['name'].encode('utf8'))
            classes.append(class_map[obj['name']])
            truncated.append(int(obj['truncated']))
            views.append(obj['pose'].encode('utf8'))

    example = tf.train.Example(features = tf.train.Features(feature = {
        'image/height': tf.train.Feature(int64_list = tf.train.Int64List(value = [height])),
        'image/width': tf.train.Feature(int64_list = tf.train.Int64List(value = [width])),
        'image/filename': tf.train.Feature(bytes_list = tf.train.BytesList(value = [annotation['filename'].encode('utf8')])),
        'image/source_id': tf.train.Feature(bytes_list = tf.train.BytesList(value = [annotation['filename'].encode('utf8')])),
        'image/key/sha256': tf.train.Feature(bytes_list = tf.train.BytesList(value = [key.encode('utf8')])),
        'image/encoded': tf.train.Feature(bytes_list = tf.train.BytesList(value = [img_raw])),
        'image/format': tf.train.Feature(bytes_list = tf.train.BytesList(value = ['jpeg'.encode('utf8')])),
```

```python
            'image/object/bbox/xmin': tf.train.Feature(float_list=tf.train.FloatList(value=xmin)),
            'image/object/bbox/xmax': tf.train.Feature(float_list=tf.train.FloatList(value=xmax)),
            'image/object/bbox/ymin': tf.train.Feature(float_list=tf.train.FloatList(value=ymin)),
            'image/object/bbox/ymax': tf.train.Feature(float_list=tf.train.FloatList(value=ymax)),
            'image/object/class/text': tf.train.Feature(bytes_list=tf.train.BytesList(value=classes_text)),
            'image/object/class/label': tf.train.Feature(int64_list=tf.train.Int64List(value=classes)),
            'image/object/difficult': tf.train.Feature(int64_list=tf.train.Int64List(value=difficult_obj)),
            'image/object/truncated': tf.train.Feature(int64_list=tf.train.Int64List(value=truncated)),
            'image/object/view': tf.train.Feature(bytes_list=tf.train.BytesList(value=views)),
        }))
    return example


def parse_xml(xml):
    if not len(xml):
        return {xml.tag: xml.text}
    result = {}
    for child in xml:
        child_result = parse_xml(child)
        if child.tag! = 'object':
            result[child.tag] = child_result[child.tag]
        else:
            if child.tag not in result:
                result[child.tag] = []
            result[child.tag].append(child_result[child.tag])
    return {xml.tag: result}
def main(_argv):
    class_map = {name: idx for idx, name in enumerate(
        open(FLAGS.classes).read().splitlines())}
    logging.info("Class mapping loaded: %s", class_map)

    writer = tf.io.TFRecordWriter(FLAGS.output_file)
    image_list = open(os.path.join(
        FLAGS.data_dir, 'ImageSets', 'Main', '%s.txt' % FLAGS.split)).read().splitlines()
    logging.info("Image list loaded:% d", len(image_list))
    for name in tqdm.tqdm(image_list):
        annotation_xml = os.path.join(
            FLAGS.data_dir, 'Annotations', name + '.xml')
        annotation_xml = lxml.etree.fromstring(open(annotation_xml).read())
```

```
            annotation = parse_xml(annotation_xml)['annotation']
            tf_example = build_example(annotation, class_map)
            writer.write(tf_example.SerializeToString())
    writer.close()
    logging.info("Done")

if __name__ == '__main__':
    app.run(main)
```

运行 voc2007_tfrecord.py 程序,分别转换 VOC2007 的训练集和验证集,在命令行中输入以下代码,运行后的转换数据如图 6.17 所示。

```
python voc2007_tfrecord.py--data_dir./datasets/VOCdevkit/VOC2007--split train --output_file./data/voc2007_train.tfrecord
```

```
python voc2007_tfrecord.py--data_dir./datasets/VOCdevkit/VOC2007--split val --output_file./data/voc2007_val.tfrecord
```

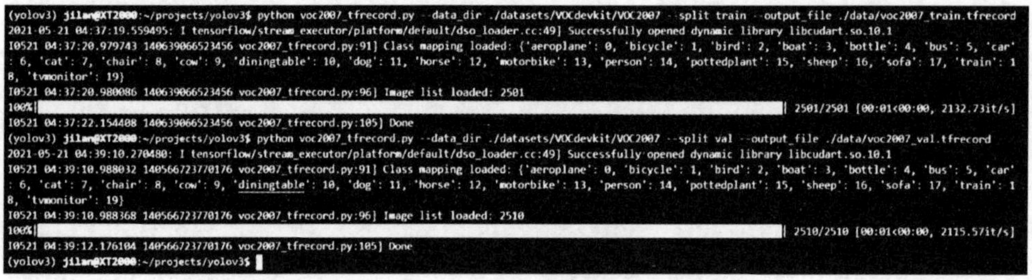

图 6.17 数据集转换

VOC2007 训练集和验证集转换完成后继续对 VOC2012 数据集进行同样的处理,将图片和 XML 标注数据转换成 TFRecord 二进制文件,在 ./data 目录下新建 voc2012.names 目标检测类别文件,在命令行中输入以下命令。

```
cd data/
touch voc2012.names
```

在 voc2012.names 中逐行写入目标检测的类别信息,VOC2007 和 VOC2012 数据集在目标检测任务中都有 20 个相同的物体类别,在代码编辑器中输入以下代码。

```
aeroplane
bicycle
bird
boat
bottle
bus
car
cat
chair
cow
diningtable
dog
```

```
horse
motorbike
person
pottedplant
sheep
sofa
train
tvmonitor
```

在工程目录下新建 voc2012_tfrecord.py，voc2012_tfrecord.py 程序将 VOC2012 数据集转换为 TFRecord 二进制格式，在代码编辑器中输入以下代码。

```
import time
import os
import hashlib

from absl import app,flags,logging
from absl.flags import FLAGS
import tensorflow as tf
import lxml.etree
import tqdm
flags.DEFINE_string('data_dir','./datasets/VOCdevkit/VOC2012/','PASCAL VOC2012 dir')
flags.DEFINE_enum('split','train',['train','val'],'train or val')
flags.DEFINE_string('output_file','./data/voc2012_train.tfrecord','outpot tfrecord dataset')
flags.DEFINE_string('classes','./data/voc2012.names','voc.names file')


def build_example(annotation,class_map):
    img_path = os.path.join(
        FLAGS.data_dir,'JPEGImages',annotation['filename'])
    img_raw = open(img_path,'rb').read()
    key = hashlib.sha256(img_raw).hexdigest()

    width = int(annotation['size']['width'])
    height = int(annotation['size']['height'])

    xmin = []
    ymin = []
    xmax = []
    ymax = []
    classes = []
    classes_text = []
    truncated = []
    views = []
    difficult_obj = []
    if 'object' in annotation:
        for obj in annotation['object']:
```

```python
                    difficult = bool(int(obj['difficult']))
                    difficult_obj.append(int(difficult))

                    xmin.append(float(obj['bndbox']['xmin']) / width)
                    ymin.append(float(obj['bndbox']['ymin']) / height)
                    xmax.append(float(obj['bndbox']['xmax']) / width)
                    ymax.append(float(obj['bndbox']['ymax']) / height)
                    classes_text.append(obj['name'].encode('utf8'))
                    classes.append(class_map[obj['name']])
                    truncated.append(int(obj['truncated']))
                    views.append(obj['pose'].encode('utf8'))

        example = tf.train.Example(features = tf.train.Features(feature = {
            'image/height': tf.train.Feature(int64_list = tf.train.Int64List(value = [height])),
            'image/width': tf.train.Feature(int64_list = tf.train.Int64List(value = [width])),
            'image/filename': tf.train.Feature(bytes_list = tf.train.BytesList(value = [annotation['filename'].encode('utf8')])),
            'image/source_id': tf.train.Feature(bytes_list = tf.train.BytesList(value = [annotation['filename'].encode('utf8')])),
            'image/key/sha256': tf.train.Feature(bytes_list = tf.train.BytesList(value = [key.encode('utf8')])),
            'image/encoded': tf.train.Feature(bytes_list = tf.train.BytesList(value = [img_raw])),
            'image/format': tf.train.Feature(bytes_list = tf.train.BytesList(value = ['jpeg'.encode('utf8')])),
            'image/object/bbox/xmin': tf.train.Feature(float_list = tf.train.FloatList(value = xmin)),
            'image/object/bbox/xmax': tf.train.Feature(float_list = tf.train.FloatList(value = xmax)),
            'image/object/bbox/ymin': tf.train.Feature(float_list = tf.train.FloatList(value = ymin)),
            'image/object/bbox/ymax': tf.train.Feature(float_list = tf.train.FloatList(value = ymax)),
            'image/object/class/text': tf.train.Feature(bytes_list = tf.train.BytesList(value = classes_text)),
            'image/object/class/label': tf.train.Feature(int64_list = tf.train.Int64List(value = classes)),
            'image/object/difficult': tf.train.Feature(int64_list = tf.train.Int64List(value = difficult_obj)),
            'image/object/truncated': tf.train.Feature(int64_list = tf.train.Int64List(value = truncated)),
            'image/object/view': tf.train.Feature(bytes_list = tf.train.BytesList(value = views)),
        }))
        return example

    def parse_xml(xml):
```

```python
        if not len(xml):
            return {xml.tag: xml.text}
        result = {}
        for child in xml:
            child_result = parse_xml(child)
            if child.tag ! = 'object':
                result[child.tag] = child_result[child.tag]
            else:
                if child.tag not in result:
                    result[child.tag] = []
                result[child.tag].append(child_result[child.tag])
        return {xml.tag: result}

    def main(_argv):
        class_map = {name: idx for idx, name in enumerate(
            open(FLAGS.classes).read().splitlines())}
        logging.info("Class mapping loaded:% s", class_map)

        writer = tf.io.TFRecordWriter(FLAGS.output_file)
        image_list = open(os.path.join(
            FLAGS.data_dir, 'ImageSets', 'Main', '% s.txt'% FLAGS.split)).read().splitlines()
        logging.info("Image list loaded:% d", len(image_list))
        for name in tqdm.tqdm(image_list):
            annotation_xml = os.path.join(
                FLAGS.data_dir, 'Annotations', name + '.xml')
            annotation_xml = lxml.etree.fromstring(open(annotation_xml).read())
            annotation = parse_xml(annotation_xml)['annotation']
            tf_example = build_example(annotation, class_map)
            writer.write(tf_example.SerializeToString())
        writer.close()
        logging.info("Done")

    if __name__ == '__main__':
        app.run(main)
```

运行 voc2012_tfrecord.py 程序，分别转换 VOC2012 的训练集和验证集，在命令行中输入以下命令，运行命令后转换数据集如图 6.18 所示。

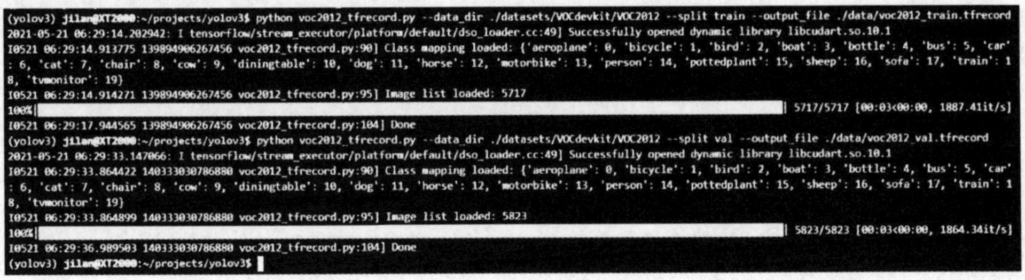

图 6.18　数据集转换

```
python voc2012_tfrecord.py--data_dir ./datasets/VOCdevkit/VOC2012--split train --
output_file ./data/voc2012_train.tfrecord

python voc2012_tfrecord.py--data_dir ./datasets/VOCdevkit/VOC2012--split val --
output_file ./data/voc2012_val.tfrecord
```

步骤3:构建数据管道。

在工程文件夹下新建文件夹yolov3,并在yolov3文件夹下新建dataset.py文件,在命令行中输入以下代码。

```
mkdir yolov3
cd yolov3/
touch dataset.py
```

文件dataset.py需要载入TFRecord文件中,并且定义训练集和验证集的数据流。在代码编辑器中输入以下代码。

```python
import tensorflow as tf
from absl.flags import FLAGS


@tf.function
def transform_targets_for_output(y_true,grid_size,anchor_idxs):
    #y_true: (N,boxes,(x1,y1,x2,y2,class,best_anchor))
    N = tf.shape(y_true)[0]

    #y_true_out: (N,grid,grid,anchors,[x1,y1,x2,y2,obj,class])
    y_true_out = tf.zeros(
        (N,grid_size,grid_size,tf.shape(anchor_idxs)[0],6))

    anchor_idxs = tf.cast(anchor_idxs,tf.int32)
    indexes = tf.TensorArray(tf.int32,1,dynamic_size = True)
    updates = tf.TensorArray(tf.float32,1,dynamic_size = True)
    idx = 0
    for i in tf.range(N):
        for j in tf.range(tf.shape(y_true)[1]):
            if tf.equal(y_true[i][j][2],0):
                continue
            anchor_eq = tf.equal(
                anchor_idxs,tf.cast(y_true[i][j][5],tf.int32))

            if tf.reduce_any(anchor_eq):
                box = y_true[i][j][0:4]
                box_xy = (y_true[i][j][0:2] + y_true[i][j][2:4]) / 2

                anchor_idx = tf.cast(tf.where(anchor_eq),tf.int32)
                grid_xy = tf.cast(box_xy // (1/grid_size),tf.int32)

                #grid[y][x][anchor] = (tx,ty,bw,bh,obj,class)
```

```python
                    indexes = indexes.write(
                        idx,[i,grid_xy[1],grid_xy[0],anchor_idx[0][0]])
                    updates = updates.write(
                        idx,[box[0],box[1],box[2],box[3],1,y_true[i][j][4]])
                    idx += 1
    return tf.tensor_scatter_nd_update(
        y_true_out,indexes.stack(),updates.stack())


def transform_targets(y_train,anchors,anchor_masks,size):
    y_outs = []
    grid_size = size //32

    #计算 GT boxes 的 anchor[index]
    anchors = tf.cast(anchors,tf.float32)
    anchor_area = anchors[...,0]* anchors[...,1]
    box_wh = y_train[...,2:4] - y_train[...,0:2]
    box_wh = tf.tile(tf.expand_dims(box_wh,-2),
                    (1,1,tf.shape(anchors)[0],1))
    box_area = box_wh[...,0]* box_wh[...,1]
    intersection = tf.minimum(box_wh[...,0],anchors[...,0])* \
        tf.minimum(box_wh[...,1],anchors[...,1])
    iou = intersection / (box_area + anchor_area - intersection)
    anchor_idx = tf.cast(tf.argmax(iou,axis =-1),tf.float32)
    anchor_idx = tf.expand_dims(anchor_idx,axis =-1)

    y_train = tf.concat([y_train,anchor_idx],axis =-1)

    for anchor_idxs in anchor_masks:
        y_outs.append(transform_targets_for_output(
            y_train,grid_size,anchor_idxs))
        grid_size *= 2

    return tuple(y_outs)

def transform_images(x_train,size):
    x_train = tf.image.resize(x_train,(size,size))
    x_train = x_train / 255
    return x_train

IMAGE_FEATURE_MAP = {
    'image/encoded': tf.io.FixedLenFeature([],tf.string),
    'image/object/bbox/xmin': tf.io.VarLenFeature(tf.float32),
    'image/object/bbox/ymin': tf.io.VarLenFeature(tf.float32),
    'image/object/bbox/xmax': tf.io.VarLenFeature(tf.float32),
    'image/object/bbox/ymax': tf.io.VarLenFeature(tf.float32),
    'image/object/class/text': tf.io.VarLenFeature(tf.string),
}
```

```python
    def parse_tfrecord(tfrecord,class_table,size):
        x = tf.io.parse_single_example(tfrecord,IMAGE_FEATURE_MAP)
        x_train = tf.image.decode_jpeg(x['image/encoded'],channels = 3)
        x_train = tf.image.resize(x_train,(size,size))
        class_text = tf.sparse.to_dense(
            x['image/object/class/text'],default_value = '')
        labels = tf.cast(class_table.lookup(class_text),tf.float32)
        y_train = tf.stack([tf.sparse.to_dense(x['image/object/bbox/xmin']),
                    tf.sparse.to_dense(x['image/object/bbox/ymin']),
                    tf.sparse.to_dense(x['image/object/bbox/xmax']),
                    tf.sparse.to_dense(x['image/object/bbox/ymax']),
                    labels],axis = 1)

        paddings = [[0,FLAGS.yolo_max_boxes-tf.shape(y_train)[0]],[0,0]]
        y_train = tf.pad(y_train,paddings)

        return x_train,y_train

    def load_tfrecord_dataset(file_pattern,class_file,size = 416):
        LINE_NUMBER = -1
        class_table = tf.lookup.StaticHashTable(tf.lookup.TextFileInitializer(
            class_file,tf.string,0,tf.int64,LINE_NUMBER,delimiter = "\n"),-1)

        files = tf.data.Dataset.list_files(file_pattern)
        dataset = files.flat_map(tf.data.TFRecordDataset)
        return dataset.map(lambda x: parse_tfrecord(x,class_table,size))

    def load_fake_dataset():
        x_train = tf.image.decode_jpeg(
            open('./data/girl.png','rb').read(),channels = 3)
        x_train = tf.expand_dims(x_train,axis = 0)

        labels = [
            [0.18494931,0.03049111,0.9435849,0.96302897,0],
            [0.01586703,0.35938117,0.17582396,0.6069674,56],
            [0.09158827,0.48252046,0.26967454,0.6403017,67]
        ] + [[0,0,0,0,0]] * 5
        y_train = tf.convert_to_tensor(labels,tf.float32)
        y_train = tf.expand_dims(y_train,axis = 0)

        return tf.data.Dataset.from_tensor_slices((x_train,y_train))
```

步骤4:构建模型。

yolov3 算法是在工业生产制造中比较流行的单阶段目标检测算法,YOLO 系列的目标检测算法在目标检测史上占有一定的地位,v3 的算法是在 v1 和 v2 的基础上逐渐创新形成的。

本项目使用的 yolov3 整体模型结构如图 6.19 所示,根据该模型可以开始构建 yolov3

的神经网络模型。

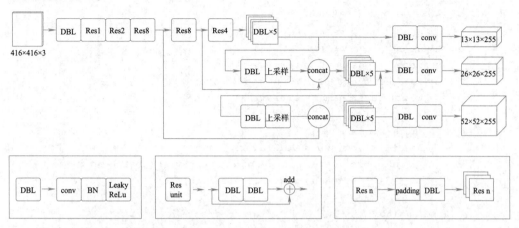

图 6.19　yolov3 模型结构图

新建 yolov3/utils.py 文件,在代码编辑器中输入以下代码。

```
from absl import logging
import numpy as np
import tensorflow as tf
import cv2

YOLOV3_LAYER_LIST = [
    'yolo_darknet',
    'yolo_conv_0',
    'yolo_output_0',
    'yolo_conv_1',
    'yolo_output_1',
    'yolo_conv_2',
    'yolo_output_2',
]

YOLOV3_TINY_LAYER_LIST = [
    'yolo_darknet',
    'yolo_conv_0',
    'yolo_output_0',
    'yolo_conv_1',
    'yolo_output_1',
]


def load_darknet_weights(model,weights_file,tiny = False):
    wf = open(weights_file,'rb')
    major,minor,revision,seen,_ = np.fromfile(wf,dtype = np.int32,count = 5)

    if tiny:
        layers = YOLOV3_TINY_LAYER_LIST
```

```python
        else:
            layers = YOLOV3_LAYER_LIST

    for layer_name in layers:
        sub_model = model.get_layer(layer_name)
        for i,layer in enumerate(sub_model.layers):
            if not layer.name.startswith('conv2d'):
                continue
            batch_norm = None
            if i + 1 < len(sub_model.layers) and \
                sub_model.layers[i + 1].name.startswith('batch_norm'):
                batch_norm = sub_model.layers[i + 1]

            logging.info("{}/{} {}".format(
                sub_model.name,layer.name,'bn' if batch_norm else 'bias'))

            filters = layer.filters
            size = layer.kernel_size[0]
            in_dim = layer.get_input_shape_at(0)[-1]

            if batch_norm is None:
                conv_bias = np.fromfile(wf,dtype = np.float32,count = filters)
            else:
                #darknet [beta,gamma,mean,variance]
                bn_weights = np.fromfile(
                    wf,dtype = np.float32,count = 4 * filters)
                #tf [gamma,beta,mean,variance]
                bn_weights = bn_weights.reshape((4,filters))[[1,0,2,3]]

            #darknet shape (out_dim,in_dim,height,width)
            conv_shape = (filters,in_dim,size,size)
            conv_weights = np.fromfile(
                wf,dtype = np.float32,count = np.product(conv_shape))
            #tf shape (height,width,in_dim,out_dim)
            conv_weights = conv_weights.reshape(
                conv_shape).transpose([2,3,1,0])

            if batch_norm is None:
                layer.set_weights([conv_weights,conv_bias])
            else:
                layer.set_weights([conv_weights])
                batch_norm.set_weights(bn_weights)

    assert len(wf.read()) == 0,'failed to read all data'
    wf.close()

def broadcast_iou(box_1,box_2):
    box_1 = tf.expand_dims(box_1,-2)
    box_2 = tf.expand_dims(box_2,0)
```

```python
        new_shape = tf.broadcast_dynamic_shape(tf.shape(box_1), tf.
shape(box_2))
        box_1 = tf.broadcast_to(box_1, new_shape)
        box_2 = tf.broadcast_to(box_2, new_shape)

        int_w = tf.maximum(tf.minimum(box_1[..., 2], box_2[..., 2]) -
                           tf.maximum(box_1[..., 0], box_2[..., 0]), 0)
        int_h = tf.maximum(tf.minimum(box_1[..., 3], box_2[..., 3]) -
                           tf.maximum(box_1[..., 1], box_2[..., 1]), 0)
        int_area = int_w * int_h
        box_1_area = (box_1[..., 2] - box_1[..., 0]) * (box_1[..., 3] - box_1[..., 1])
        box_2_area = (box_2[..., 2] - box_2[..., 0]) * (box_2[..., 3] - box_2[..., 1])
        return int_area / (box_1_area + box_2_area - int_area)

    def draw_outputs(img, outputs, class_names):
        boxes, objectness, classes, nums = outputs
        boxes, objectness, classes, nums = boxes[0], objectness[0], classes[0], nums[0]
        wh = np.flip(img.shape[0:2])
        for i in range(nums):
            x1y1 = tuple((np.array(boxes[i][0:2]) * wh).astype(np.int32))
            x2y2 = tuple((np.array(boxes[i][2:4]) * wh).astype(np.int32))
            img = cv2.rectangle(img, x1y1, x2y2, (255, 0, 0), 2)
            img = cv2.putText(img, '{} {:.4f}'.format(
                class_names[int(classes[i])], objectness[i]),
                x1y1, cv2.FONT_HERSHEY_COMPLEX_SMALL, 1, (0, 0, 255), 2)
        return img

    def draw_labels(x, y, class_names):
        img = x.numpy()
        boxes, classes = tf.split(y, (4, 1), axis=-1)
        classes = classes[..., 0]
        wh = np.flip(img.shape[0:2])
        for i in range(len(boxes)):
            x1y1 = tuple((np.array(boxes[i][0:2]) * wh).astype(np.int32))
            x2y2 = tuple((np.array(boxes[i][2:4]) * wh).astype(np.int32))
            img = cv2.rectangle(img, x1y1, x2y2, (255, 0, 0), 2)
            img = cv2.putText(img, class_names[classes[i]],
                x1y1, cv2.FONT_HERSHEY_COMPLEX_SMALL, 1, (0, 0, 255), 2)
        return img

    def freeze_all(model, frozen=True):
        model.trainable = not frozen
        if isinstance(model, tf.keras.Model):
            for l in model.layers:
                freeze_all(l, frozen)
```

新建 yolov3/models.py 文件,在代码编辑器中输入以下代码。

```python
from absl import flags
from absl.flags import FLAGS
```

```python
import numpy as np
import tensorflow as tf
from tensorflow.keras import Model
from tensorflow.keras.layers import (
    Add,
    Concatenate,
    Conv2D,
    Input,
    Lambda,
    LeakyReLU,
    MaxPool2D,
    UpSampling2D,
    ZeroPadding2D,
    BatchNormalization,
)
from tensorflow.keras.regularizers import l2
from tensorflow.keras.losses import (
    binary_crossentropy,
    sparse_categorical_crossentropy
)
from .utils import broadcast_iou

flags.DEFINE_integer('yolo_max_boxes',100,
                    'maximum number of boxes per image')
flags.DEFINE_float('yolo_iou_threshold',0.5,'iou threshold')
flags.DEFINE_float('yolo_score_threshold',0.5,'score threshold')

yolo_anchors = np.array([(10,13),(16,30),(33,23),(30,61),(62,45),
            (59,119),(116,90),(156,198),(373,326)],np.float32) / 416
yolo_anchor_masks = np.array([[6,7,8],[3,4,5],[0,1,2]])

yolo_tiny_anchors = np.array([(10,14),(23,27),(37,58),
            (81,82),(135,169),  (344,319)],np.float32) / 416
yolo_tiny_anchor_masks = np.array([[3,4,5],[0,1,2]])
def DarknetConv(x,filters,size,strides=1,batch_norm=True):
    if strides == 1:
        padding = 'same'
    else:
        x = ZeroPadding2D(((1,0),(1,0)))(x)   #top left half-padding
        padding = 'valid'
    x = Conv2D(filters=filters,kernel_size=size,
            strides=strides,padding=padding,
            use_bias=not batch_norm,kernel_regularizer=l2(0.0005))(x)
    if batch_norm:
        x = BatchNormalization()(x)
        x = LeakyReLU(alpha=0.1)(x)
    return x
```

```python
def DarknetResidual(x,filters):
    prev = x
    x = DarknetConv(x,filters //2,1)
    x = DarknetConv(x,filters,3)
    x = Add()([prev,x])
    return x

def DarknetBlock(x,filters,blocks):
    x = DarknetConv(x,filters,3,strides = 2)
    for _ in range(blocks):
        x = DarknetResidual(x,filters)
    return x

def Darknet(name = None):
    x = inputs = Input([None,None,3])
    x = DarknetConv(x,32,3)
    x = DarknetBlock(x,64,1)
    x = DarknetBlock(x,128,2)              #skip connection
    x = x_36 = DarknetBlock(x,256,8)       #skip connection
    x = x_61 = DarknetBlock(x,512,8)
    x = DarknetBlock(x,1024,4)
    return tf.keras.Model(inputs,(x_36,x_61,x),name = name)
def DarknetTiny(name = None):
    x = inputs = Input([None,None,3])
    x = DarknetConv(x,16,3)
    x = MaxPool2D(2,2,'same')(x)
    x = DarknetConv(x,32,3)
    x = MaxPool2D(2,2,'same')(x)
    x = DarknetConv(x,64,3)
    x = MaxPool2D(2,2,'same')(x)
    x = DarknetConv(x,128,3)
    x = MaxPool2D(2,2,'same')(x)
    x = x_8 = DarknetConv(x,256,3)    #skip connection
    x = MaxPool2D(2,2,'same')(x)
    x = DarknetConv(x,512,3)
    x = MaxPool2D(2,1,'same')(x)
    x = DarknetConv(x,1024,3)
    return tf.keras.Model(inputs,(x_8,x),name = name)

def YoloConv(filters,name = None):
    def yolo_conv(x_in):
        if isinstance(x_in,tuple):
            inputs = Input(x_in[0].shape[1:]),Input(x_in[1].shape[1:])

            x,x_skip = inputs

            #concat with skip connection
            x = DarknetConv(x,filters,1)
            x = UpSampling2D(2)(x)
```

```
                x = Concatenate()([x,x_skip])
            else:
                x = inputs = Input(x_in.shape[1:])

        x = DarknetConv(x,filters,1)
        x = DarknetConv(x,filters* 2,3)
        x = DarknetConv(x,filters,1)
        x = DarknetConv(x,filters* 2,3)
        x = DarknetConv(x,filters,1)
        return Model(inputs,x,name = name)(x_in)
    return yolo_conv
def YoloConvTiny(filters,name = None):
    def yolo_conv(x_in):
        if isinstance(x_in,tuple):
            inputs = Input(x_in[0].shape[1:]),Input(x_in[1].shape[1:])
            x,x_skip = inputs

            #concat with skip connection
            x = DarknetConv(x,filters,1)
            x = UpSampling2D(2)(x)
            x = Concatenate()([x,x_skip])
        else:
            x = inputs = Input(x_in.shape[1:])
            x = DarknetConv(x,filters,1)

        return Model(inputs,x,name = name)(x_in)
    return yolo_conv


def YoloOutput(filters,anchors,classes,name = None):
    def yolo_output(x_in):
        x = inputs = Input(x_in.shape[1:])
        x = DarknetConv(x,filters* 2,3)
        x = DarknetConv(x,anchors* (classes + 5),1,batch_norm = False)
        x = Lambda(lambda x: tf.reshape(x, (-1,tf.shape(x)[1],tf.shape(x)[2],anchors,classes + 5)))(x)
        return tf.keras.Model(inputs,x,name = name)(x_in)
    return yolo_output


#As tensorflow lite doesn't support tf.size used in tf.meshgrid,
#we reimplemented a simple meshgrid function that use basic tf function.
def _meshgrid(n_a,n_b):

    return [
        tf.reshape(tf.tile(tf.range(n_a),[n_b]),(n_b,n_a)),
        tf.reshape(tf.repeat(tf.range(n_b),n_a),(n_b,n_a))
    ]
```

```python
def yolo_boxes(pred,anchors,classes):
    #pred: (batch_size,grid,grid,anchors,(x,y,w,h,obj,... classes))
    grid_size = tf.shape(pred)[1:3]
    box_xy,box_wh,objectness,class_probs = tf.split(
        pred,(2,2,1,classes),axis=-1)
    
    box_xy = tf.sigmoid(box_xy)
    objectness = tf.sigmoid(objectness)
    class_probs = tf.sigmoid(class_probs)
    pred_box = tf.concat((box_xy,box_wh),axis=-1)
    
    #!!! grid[x][y] == (y,x)
    grid = _meshgrid(grid_size[1],grid_size[0])
    grid = tf.expand_dims(tf.stack(grid,axis=-1),axis=2)   #[gx,gy,1,2]
    
    box_xy = (box_xy + tf.cast(grid,tf.float32)) / \
        tf.cast(grid_size,tf.float32)
    box_wh = tf.exp(box_wh) * anchors
    
    box_x1y1 = box_xy - box_wh / 2
    box_x2y2 = box_xy + box_wh / 2
    bbox = tf.concat([box_x1y1,box_x2y2],axis=-1)
    
    return bbox,objectness,class_probs,pred_box

def yolo_nms(outputs,anchors,masks,classes):
    b,c,t = [],[],[]
    
    for o in outputs:
        b.append(tf.reshape(o[0],(tf.shape(o[0])[0],-1,tf.shape(o[0])[-1])))
        c.append(tf.reshape(o[1],(tf.shape(o[1])[0],-1,tf.shape(o[1])[-1])))
        t.append(tf.reshape(o[2],(tf.shape(o[2])[0],-1,tf.shape(o[2])[-1])))
    bbox = tf.concat(b,axis=1)
    confidence = tf.concat(c,axis=1)
    class_probs = tf.concat(t,axis=1)
    
    scores = confidence* class_probs
    
    dscores = tf.squeeze(scores,axis=0)
    scores = tf.reduce_max(dscores,[1])
    bbox = tf.reshape(bbox,(-1,4))
    classes = tf.argmax(dscores,1)
    selected_indices,selected_scores = tf.image.non_max_suppression_with_scores(
        boxes=bbox,
        scores=scores,
        max_output_size=FLAGS.yolo_max_boxes,
        iou_threshold=FLAGS.yolo_iou_threshold,
```

```python
            score_threshold=FLAGS.yolo_score_threshold,
            soft_nms_sigma=0.5
        )

        num_valid_nms_boxes = tf.shape(selected_indices)[0]

        selected_indices = tf.concat([selected_indices,tf.zeros(FLAGS.yolo_max_
boxes-num_valid_nms_boxes,tf.int32)],0)
        selected_scores = tf.concat([selected_scores,tf.zeros(FLAGS.yolo_max_
boxes-num_valid_nms_boxes,tf.float32)],-1)

        boxes = tf.gather(bbox,selected_indices)
        boxes = tf.expand_dims(boxes,axis=0)
        scores = selected_scores
        scores = tf.expand_dims(scores,axis=0)
        classes = tf.gather(classes,selected_indices)
        classes = tf.expand_dims(classes,axis=0)
        valid_detections = num_valid_nms_boxes
        valid_detections = tf.expand_dims(valid_detections,axis=0)

        return boxes,scores,classes,valid_detections
    def YoloV3(size=None,channels=3,anchors=yolo_anchors,masks=yolo_anchor_
masks,classes=80,training=False):
        x = inputs = Input([size,size,channels],name='input')

        x_36,x_61,x = Darknet(name='yolo_darknet')(x)

        x = YoloConv(512,name='yolo_conv_0')(x)
        output_0 = YoloOutput(512,len(masks[0]),classes,name='yolo_output_0')(x)

        x = YoloConv(256,name='yolo_conv_1')((x,x_61))
        output_1 = YoloOutput(256,len(masks[1]),classes,name='yolo_output_1')(x)

        x = YoloConv(128,name='yolo_conv_2')((x,x_36))
        output_2 = YoloOutput(128,len(masks[2]),classes,name='yolo_output_2')(x)

        if training:
            return Model(inputs,(output_0,output_1,output_2),name='yolov3')

        boxes_0 = Lambda(lambda x: yolo_boxes(x,anchors[masks[0]],classes),name=
'yolo_boxes_0')(output_0)
        boxes_1 = Lambda(lambda x: yolo_boxes(x,anchors[masks[1]],classes),name=
'yolo_boxes_1')(output_1)
        boxes_2 = Lambda(lambda x: yolo_boxes(x,anchors[masks[2]],classes),name=
'yolo_boxes_2')(output_2)

        outputs = Lambda(lambda x: yolo_nms(x,anchors,masks,classes),name='yolo_
nms')((boxes_0[:3],boxes_1[:3],boxes_2[:3]))
```

```
        return Model(inputs,outputs,name='yolov3')
    def YoloV3Tiny(size=None,channels=3,anchors=yolo_tiny_anchors,masks=yolo
_tiny_anchor_masks,classes=80,training=False):
        x=inputs=Input([size,size,channels],name='input')

        x_8,x=DarknetTiny(name='yolo_darknet')(x)

        x=YoloConvTiny(256,name='yolo_conv_0')(x)
        output_0=YoloOutput(256,len(masks[0]),classes,name='yolo_output_0')(x)

        x=YoloConvTiny(128,name='yolo_conv_1')((x,x_8))
        output_1=YoloOutput(128,len(masks[1]),classes,name='yolo_output_1')(x)

        if training:
            return Model(inputs,(output_0,output_1),name='yolov3')

        boxes_0=Lambda(lambda x: yolo_boxes(x,anchors[masks[0]],classes),name=
'yolo_boxes_0')(output_0)
        boxes_1=Lambda(lambda x: yolo_boxes(x,anchors[masks[1]],classes),name=
'yolo_boxes_1')(output_1)
        outputs=Lambda(lambda x: yolo_nms(x,anchors,masks,classes),
                    name='yolo_nms')((boxes_0[:3],boxes_1[:3]))
        return Model(inputs,outputs,name='yolov3_tiny')

    def YoloLoss(anchors,classes=80,ignore_thresh=0.5):
        def yolo_loss(y_true,y_pred):
            #1. 转换全部预测输出
            #y_pred: (batch_size,grid,grid,anchors,(x,y,w,h,obj,...cls))
            pred_box,pred_obj,pred_class,pred_xywh=yolo_boxes(
                y_pred,anchors,classes)
            pred_xy=pred_xywh[...,0:2]
            pred_wh=pred_xywh[...,2:4]
            #2. 转换全部真实输出
            #y_true: (batch_size,grid,grid,anchors,(x1,y1,x2,y2,obj,cls))
            true_box,true_obj,true_class_idx=tf.split(
                y_true,(4,1,1),axis=-1)
            true_xy=(true_box[...,0:2]+true_box[...,2:4]) / 2
            true_wh=true_box[...,2:4]-true_box[...,0:2]

            #give higher weights to small boxes
            box_loss_scale=2-true_wh[...,0]* true_wh[...,1]

            #3. inverting the pred box equations
            grid_size=tf.shape(y_true)[1]
            grid=tf.meshgrid(tf.range(grid_size),tf.range(grid_size))
            grid=tf.expand_dims(tf.stack(grid,axis=-1),axis=2)
            true_xy=true_xy* tf.cast(grid_size,tf.float32) - \
                tf.cast(grid,tf.float32)
```

```python
            true_wh = tf.math.log(true_wh / anchors)
            true_wh = tf.where(tf.math.is_inf(true_wh),
                    tf.zeros_like(true_wh),true_wh)

        #4. calculate all masks
        obj_mask = tf.squeeze(true_obj,-1)
        #ignore false positive when iou is over threshold
        best_iou = tf.map_fn(
            lambda x: tf.reduce_max(broadcast_iou(x[0],tf.boolean_mask(
                x[1],tf.cast(x[2],tf.bool))),axis =-1),
            (pred_box,true_box,obj_mask),
            tf.float32)
        ignore_mask = tf.cast(best_iou < ignore_thresh,tf.float32)

        #5. 计算总损失
        xy_loss = obj_mask* box_loss_scale* \
            tf.reduce_sum(tf.square(true_xy - pred_xy),axis =-1)
        wh_loss = obj_mask* box_loss_scale* \
            tf.reduce_sum(tf.square(true_wh - pred_wh),axis =-1)
        obj_loss = binary_crossentropy(true_obj,pred_obj)
        obj_loss = obj_mask* obj_loss + \
            (1-obj_mask)* ignore_mask* obj_loss
        #也可以使用 binary_crossentropy 损失
        class_loss = obj_mask* sparse_categorical_crossentropy(
            true_class_idx,pred_class)

        #6. 计算总和 (batch,gridx,gridy,anchors) = > (batch,1)
        xy_loss = tf.reduce_sum(xy_loss,axis = (1,2,3))
        wh_loss = tf.reduce_sum(wh_loss,axis = (1,2,3))
        obj_loss = tf.reduce_sum(obj_loss,axis = (1,2,3))
        class_loss = tf.reduce_sum(class_loss,axis = (1,2,3))

        return xy_loss + wh_loss + obj_loss + class_loss
    return yolo_loss
```

模型文件完成后,为了便于使用建立好的模型文件,在整个 yolov3 的虚拟环境中构建 yolov3 的全局模块,这时需要在 models.py 平行的文件目录下建立_init_空文件。在命令行中输入以下代码。

```
cd yolov3
touch __init__.py
cd..
```

返回到工程文件目录下,新建 setup.py 文件构建全局模块,在命令行中输入以下代码。

```
touch setup.py
```

在 setup.py 中使用 setuptool 为前面写好的 yolov3 文件构建全局模块,模块名为 yolov3,在代码编辑器中输入以下代码。

```
from setuptools import setup

setup(name = 'yolov3',
      version = '1.0',
      packages = ['yolov3'])
```

运行 setup.py 文件为 yolov3 构建模块并安装到全局虚拟环境 yolov3 中,在命令行中输入以下代码。

```
python setup.py develop
```

测试前面的 yolov3 模块是否成功安装到全局虚拟环境中,在命令行中输入以下代码。

```
python
#进入交互环境
Python 3.7.10 (default,Feb 26 2021,18:47:35)
[GCC 7.3.0] :: Anaconda,Inc. on linux
Type "help","copyright","credits" or "license" for more information.
>>> import yolov3
>>>
#可以成功导入 yolov3 模块则表示该模块成功安装到全局虚拟环境中
```

步骤 5:训练模型。

在工程文件夹下新建训练脚本程序 train.py,打开 train.py,在代码编辑器中输入以下代码。

```
from absl import app,flags,logging
from absl.flags import FLAGS

import tensorflow as tf
import numpy as np
import cv2
from tensorflow.keras.callbacks import (
    ReduceLROnPlateau,
    EarlyStopping,
    ModelCheckpoint,
    TensorBoard
)
from yolov3.models import (
    YoloV3,YoloV3Tiny,YoloLoss,
    yolo_anchors,yolo_anchor_masks,
    yolo_tiny_anchors,yolo_tiny_anchor_masks
)
from yolov3.utils import freeze_all
import yolov3.dataset as dataset

flags.DEFINE_string('dataset','','path to dataset')
flags.DEFINE_string('val_dataset','','path to validation dataset')
flags.DEFINE_boolean('tiny',False,'yolov3 or yolov3-tiny')
flags.DEFINE_string('weights','./checkpoints/yolov3','path to weights file')
flags.DEFINE_string('classes','./data/coco.names','path to classes file')
```

```python
    flags.DEFINE_enum('mode','fit',['fit','eager_fit','eager_tf'],
                      'fit: model.fit,'
                      'eager_fit: model.fit(run_eagerly=True),'
                      'eager_tf: custom GradientTape')
    flags.DEFINE_enum('transfer','none',
                      ['none','darknet','no_output','frozen','fine_tune'],
                      'none: Training from scratch,'
                      'darknet: Transfer darknet,'
                      'no_output: Transfer all but output,'
                      'frozen: Transfer and freeze all,'
                      'fine_tune: Transfer all and freeze darknet only')
flags.DEFINE_integer('size',416,'image size')
flags.DEFINE_integer('epochs',2,'number of epochs')
flags.DEFINE_integer('batch_size',8,'batch size')
flags.DEFINE_float('learning_rate',1e-3,'learning rate')
flags.DEFINE_integer('num_classes',80,'number of classes in the model')
flags.DEFINE_integer('weights_num_classes',None,'specify num class for '
weights' file if different,'
                     'useful in transfer learning with different number of classes')


    def main(_argv):
        physical_devices=tf.config.experimental.list_physical_devices('GPU')
        for physical_device in physical_devices:
            tf.config.experimental.set_memory_growth(physical_device,True)

        if FLAGS.tiny:
            model=YoloV3Tiny(FLAGS.size,training=True,
                             classes=FLAGS.num_classes)
            anchors=yolo_tiny_anchors
            anchor_masks=yolo_tiny_anchor_masks
        else:
            model=YoloV3(FLAGS.size,training=True,classes=FLAGS.num_classes)
            anchors=yolo_anchors
            anchor_masks=yolo_anchor_masks

        if FLAGS.dataset:
            train_dataset=dataset.load_tfrecord_dataset(
                FLAGS.dataset,FLAGS.classes,FLAGS.size)
        else:
            train_dataset=dataset.load_fake_dataset()
    train_dataset=train_dataset.shuffle(buffer_size=512)
    train_dataset=train_dataset.batch(FLAGS.batch_size)
    train_dataset=train_dataset.map(lambda x,y: (
        dataset.transform_images(x,FLAGS.size),
        dataset.transform_targets(y,anchors,anchor_masks,FLAGS.size)))
    train_dataset=train_dataset.prefetch(
        buffer_size=tf.data.experimental.AUTOTUNE)
```

```
        if FLAGS.val_dataset:
            val_dataset = dataset.load_tfrecord_dataset(
                FLAGS.val_dataset,FLAGS.classes,FLAGS.size)
        else:
            val_dataset = dataset.load_fake_dataset()
        val_dataset = val_dataset.batch(FLAGS.batch_size)
        val_dataset = val_dataset.map(lambda x,y: (
            dataset.transform_images(x,FLAGS.size),
            dataset.transform_targets(y,anchors,anchor_masks,FLAGS.size)))

        #Configure the model for transfer learning
        if FLAGS.transfer == 'none':
            pass
        elif FLAGS.transfer in ['darknet','no_output']:
            #reset top layers
            if FLAGS.tiny:
                model_pretrained = YoloV3Tiny(
                    FLAGS.size,training = True,classes = FLAGS.weights_num_classes
or FLAGS.num_classes)
            else:
                model_pretrained = YoloV3(
                    FLAGS.size,training = True,classes = FLAGS.weights_num_classes
or FLAGS.num_classes)
            model_pretrained.load_weights(FLAGS.weights)

            if FLAGS.transfer == 'darknet':
                model.get_layer('yolo_darknet').set_weights(
                    model_pretrained.get_layer('yolo_darknet').get_weights())
                freeze_all(model.get_layer('yolo_darknet'))

            elif FLAGS.transfer == 'no_output':
                for l in model.layers:
                    if not l.name.startswith('yolo_output'):
                        l.set_weights(model_pretrained.get_layer(
                            l.name).get_weights())
                        freeze_all(l)

        else:
            #All other transfer require matching classes
            model.load_weights(FLAGS.weights)
            if FLAGS.transfer == 'fine_tune':
                #freeze darknet and fine tune other layers
                darknet = model.get_layer('yolo_darknet')
                freeze_all(darknet)
            elif FLAGS.transfer == 'frozen':
                #freeze everything
                freeze_all(model)

        optimizer = tf.keras.optimizers.Adam(lr = FLAGS.learning_rate)
```

```python
        loss = [YoloLoss(anchors[mask], classes = FLAGS.num_classes)
            for mask in anchor_masks]

        if FLAGS.mode = = 'eager_tf':
            #Eager mode is great for debugging
            #Non eager graph mode is recommended for real training
            avg_loss = tf.keras.metrics.Mean('loss', dtype = tf.float32)
            avg_val_loss = tf.keras.metrics.Mean('val_loss', dtype = tf.float32)

            for epoch in range(1, FLAGS.epochs + 1):
                for batch, (images, labels) in enumerate(train_dataset):
                    with tf.GradientTape() as tape:
                        outputs = model(images, training = True)
                        regularization_loss = tf.reduce_sum(model.losses)
                        pred_loss = []
                        for output, label, loss_fn in zip(outputs, labels, loss):
                            pred_loss.append(loss_fn(label, output))
                        total_loss = tf.reduce_sum(pred_loss) + regularization_loss
                    grads = tape.gradient(total_loss, model.trainable_variables)
                    optimizer.apply_gradients(
                        zip(grads, model.trainable_variables))

                    logging.info("{}_train_{},{},{}".format(
                        epoch, batch, total_loss.numpy(),
                        list(map(lambda x: np.sum(x.numpy()), pred_loss))))
                    avg_loss.update_state(total_loss)

                for batch, (images, labels) in enumerate(val_dataset):
                    outputs = model(images)
                    regularization_loss = tf.reduce_sum(model.losses)
                    pred_loss = []
                    for output, label, loss_fn in zip(outputs, labels, loss):
                        pred_loss.append(loss_fn(label, output))
                    total_loss = tf.reduce_sum(pred_loss) + regularization_loss

                    logging.info("{}_val_{},{},{}".format(
                        epoch, batch, total_loss.numpy(),
                        list(map(lambda x: np.sum(x.numpy()), pred_loss))))
                    avg_val_loss.update_state(total_loss)

                logging.info("{}, train: {}, val: {}".format(
                    epoch,
                    avg_loss.result().numpy(),
                    avg_val_loss.result().numpy()))

                avg_loss.reset_states()
                avg_val_loss.reset_states()
                model.save_weights(
                    'checkpoints/yolov3_train_{}'.format(epoch))
```

```
    else:
        model.compile(optimizer = optimizer, loss = loss,
                      run_eagerly = (FLAGS.mode = = 'eager_fit'))

        callbacks = [
            ReduceLROnPlateau(verbose = 1),
            EarlyStopping(patience = 3, verbose = 1),
            ModelCheckpoint('checkpoints/yolov3_train_{epoch}',
                            verbose = 1, save_weights_only = True),
            TensorBoard(log_dir = 'logs')
        ]

        history = model.fit(train_dataset,
                            epochs = FLAGS.epochs,
                            callbacks = callbacks,
                            validation_data = val_dataset)

if __name__ = = '__main__':
    try:
        app.run(main)
    except SystemExit:
        pass
```

训练程序 train.py 完成后,需要为 yolov3 准备一个预训练文件,一般的模型都需要准备预训练模型,预训练模型一般是在庞大的 ImageNet 数据集上进行初始化训练的模型参数。当然也可以从零开始训练,但是因为自选训练集的局限性,从零开始训练的模型参数极有可能不收敛。因此,有必要准备预训练模型。这里选择 yolov3 官方从 C 源代码训练出的 yolov3.weights 模型参数作为预训练模型。首先从资源服务器下载预训练模型,在命令行中输入以下代码。下载预训练模型参数如图 6.20 所示。

```
wget https://172.16.33.72/dataset/yolov3.weights-O data/yolov3.weights
```

图 6.20　下载预训练模型参数

预训练模型准备好后,data/yolov3.weights 是 C 语言源代码格式的保存形式,还需要进一步将预训练模型转换成 TensorFlow 的 checkpoint 模型格式。在工程文件夹下创建 convert.py,在代码编辑器中输入以下代码。

```
from absl import app, flags, logging
from absl.flags import FLAGS
import numpy as np
from yolov3.models import YoloV3, YoloV3Tiny
```

```python
from yolov3.utils import load_darknet_weights
import tensorflow as tf

flags.DEFINE_string('weights','./data/yolov3.weights','path to weights file')
flags.DEFINE_string('output','./checkpoints/yolov3','path to output')
flags.DEFINE_boolean('tiny',False,'yolov3 or yolov3-tiny')
flags.DEFINE_integer('num_classes',80,'number of classes in the model')

def main(_argv):
    physical_devices = tf.config.experimental.list_physical_devices('GPU')
    if len(physical_devices) > 0:
        tf.config.experimental.set_memory_growth(physical_devices[0],True)

    if FLAGS.tiny:
        yolo = YoloV3Tiny(classes = FLAGS.num_classes)
    else:
        yolo = YoloV3(classes = FLAGS.num_classes)
    yolo.summary()
    logging.info('model created')

    load_darknet_weights(yolo,FLAGS.weights,FLAGS.tiny)
    logging.info('weights loaded')

    img = np.random.random((1,320,320,3)).astype(np.float32)
    output = yolo(img)
    logging.info('sanity check passed')

    yolo.save_weights(FLAGS.output)
    logging.info('weights saved')

if __name__ == '__main__':
    try:
        app.run(main)
    except SystemExit:
        pass
```

完成 convert.py 后运行,转换 yolov3.weights 为 checkpoints 文件。在命令行中输入如下代码,运行后转换参数如图 6.21 所示。

```
python convert.py
```

完成后工程文件夹下会出现一个 checkpoint 文件夹,预训练模型 checkpoints/yolov3 如下。

```
checkpoints/
├─checkpoint
├─yolov3.data-00000-of-00001
└─yolov3.index

0 directories,3 files
```

```
Model: "yolov3"
__________________________________________________________________________________________________
Layer (type)                    Output Shape         Param #     Connected to
==================================================================================================
input (InputLayer)              [(None, None, None,  0
__________________________________________________________________________________________________
yolo_darknet (Functional)       ((None, None, None,  40620640    input[0][0]
__________________________________________________________________________________________________
yolo_conv_0 (Functional)        (None, None, None, 5 11024384    yolo_darknet[0][2]
__________________________________________________________________________________________________
yolo_conv_1 (Functional)        (None, None, None, 2 2957312     yolo_conv_0[0][0]
                                                                 yolo_darknet[0][1]
__________________________________________________________________________________________________
yolo_conv_2 (Functional)        (None, None, None, 1 741376      yolo_conv_1[0][0]
                                                                 yolo_darknet[0][0]
__________________________________________________________________________________________________
yolo_output_0 (Functional)      (None, None, None, 3 4984063     yolo_conv_0[0][0]
__________________________________________________________________________________________________
yolo_output_1 (Functional)      (None, None, None, 3 1312511     yolo_conv_1[0][0]
__________________________________________________________________________________________________
yolo_output_2 (Functional)      (None, None, None, 3 361471      yolo_conv_2[0][0]
__________________________________________________________________________________________________
yolo_boxes_0 (Lambda)           ((None, None, None,  0           yolo_output_0[0][0]
__________________________________________________________________________________________________
yolo_boxes_1 (Lambda)           ((None, None, None,  0           yolo_output_1[0][0]
__________________________________________________________________________________________________
yolo_boxes_2 (Lambda)           ((None, None, None,  0           yolo_output_2[0][0]
__________________________________________________________________________________________________
yolo_nms (Lambda)               ((1, None, 4), (1, N 0           yolo_boxes_0[0][0]
                                                                 yolo_boxes_0[0][1]
                                                                 yolo_boxes_0[0][2]
                                                                 yolo_boxes_1[0][0]
                                                                 yolo_boxes_1[0][1]
                                                                 yolo_boxes_1[0][2]
                                                                 yolo_boxes_2[0][0]
                                                                 yolo_boxes_2[0][1]
                                                                 yolo_boxes_2[0][2]
==================================================================================================
Total params: 62,001,757
Trainable params: 61,949,149
Non-trainable params: 52,608
```

图 6.21　转换预训练参数

预训练模型准备好后就可以开始训练了，下面使用 VOC2007 的训练集训练 20 类的目标检测模型。在命令行中输入以下代码。运行后的训练模型如图 6.22 所示。

```
python train.py--dataset./data/voc2007_train.tfrecord--val_dataset./data/voc2007_val.tfrecord--classes./data/voc2007.names--num_classes 20--mode fit--transfer darknet--batch_size 16--epochs 10--weights./checkpoints/yolov3--weights_num_classes 80
```

```
Dumped tool data for overview_page.pb to logs/train/plugins/profile/2021_05_24_04_01_50/XT2000.overview_page.pb
Dumped tool data for input_pipeline.pb to logs/train/plugins/profile/2021_05_24_04_01_50/XT2000.input_pipeline.pb
Dumped tool data for tensorflow_stats.pb to logs/train/plugins/profile/2021_05_24_04_01_50/XT2000.tensorflow_stats.pb
Dumped tool data for kernel_stats.pb to logs/train/plugins/profile/2021_05_24_04_01_50/XT2000.kernel_stats.pb

157/157 [==============================] - 1667s 11s/step - loss: 1916.9433 - yolo_output_0_loss: 116.6237 - yolo_output_1_loss: 358.6729 - yolo_output_2_loss: 1430.6831 - val_loss: 187.4308 - val_yolo_output_0_loss: 14.1617 - val_yolo_output_1_loss: 23.9285 - val_yolo_output_2_loss: 138.2619

Epoch 00001: saving model to checkpoints/yolov3_train_1.tf
Epoch 2/10
157/157 [==============================] - 1682s 11s/step - loss: 72.7454 - yolo_output_0_loss: 9.9887 - yolo_output_1_loss: 11.7052 - yolo_output_2_loss: 39.9724 - val_loss: 57.0493 - val_yolo_output_0_loss: 7.4971 - val_yolo_output_1_loss: 10.5690 - val_yolo_output_2_loss: 27.9036

Epoch 00002: saving model to checkpoints/yolov3_train_2.tf
Epoch 3/10
157/157 [==============================] - 1682s 11s/step - loss: 43.7605 - yolo_output_0_loss: 6.6647 - yolo_output_1_loss: 8.5873 - yolo_output_2_loss: 17.4280 - val_loss: 39.3536 - val_yolo_output_0_loss: 6.2709 - val_yolo_output_1_loss: 8.7276 - val_yolo_output_2_loss: 13.2712

Epoch 00003: saving model to checkpoints/yolov3_train_3.tf
Epoch 4/10
157/157 [==============================] - 1680s 11s/step - loss: 33.5898 - yolo_output_0_loss: 5.0831 - yolo_output_1_loss: 6.9831 - yolo_output_2_loss: 10.4381 - val_loss: 33.6217 - val_yolo_output_0_loss: 5.9951 - val_yolo_output_1_loss: 7.4870 - val_yolo_output_2_loss: 9.0486

Epoch 00004: saving model to checkpoints/yolov3_train_4.tf
Epoch 5/10
157/157 [==============================] - 1685s 11s/step - loss: 29.0241 - yolo_output_0_loss: 4.2222 - yolo_output_1_loss: 6.0717 - yolo_output_2_loss: 7.6380 - val_loss: 30.2372 - val_yolo_output_0_loss: 5.7946 - val_yolo_output_1_loss: 6.7093 - val_yolo_output_2_loss: 6.6373

Epoch 00005: saving model to checkpoints/yolov3_train_5.tf
Epoch 6/10
157/157 [==============================] - 1686s 11s/step - loss: 25.1834 - yolo_output_0_loss: 2.9244 - yolo_output_1_loss: 5.2336 - yolo_output_2_loss: 5.9286 - val_loss: 29.0202 - val_yolo_output_0_loss: 5.3876 - val_yolo_output_1_loss: 7.0325 - val_yolo_output_2_loss: 5.4992

Epoch 00006: saving model to checkpoints/yolov3_train_6.tf
Epoch 7/10
157/157 [==============================] - 1690s 11s/step - loss: 22.7937 - yolo_output_0_loss: 2.1598 - yolo_output_1_loss: 4.6015 - yolo_output_2_loss: 4.9315 - val_loss: 27.8155 - val_yolo_output_0_loss: 5.3849 - val_yolo_output_1_loss: 6.4913 - val_yolo_output_2_loss: 4.8347
```

图 6.22　训练模型

保存的模型为 TensorFlow 内置的 checkpoint 模型,保存模型在 ./checkpoints 目录中,checkpoints 目录结构如下。

```
checkpoints/
├─checkpoint
├─yolov3.data-00000-of-00001
├─yolov3.index
├─yolov3_train_1.data-00000-of-00001
├─yolov3_train_1.index
├─yolov3_train_10.data-00000-of-00001
├─yolov3_train_10.index
├─yolov3_train_2.data-00000-of-00001
├─yolov3_train_2.index
├─yolov3_train_3.data-00000-of-00001
├─yolov3_train_3.index
├─yolov3_train_4.data-00000-of-00001
├─yolov3_train_4.index
├─yolov3_train_5.data-00000-of-00001
├─yolov3_train_5.index
├─yolov3_train_6.data-00000-of-00001
├─yolov3_train_6.index
├─yolov3_train_7.data-00000-of-00001
├─yolov3_train_7.index
├─yolov3_train_8.data-00000-of-00001
├─yolov3_train_8.index
├─yolov3_train_9.data-00000-of-00001
└─yolov3_train_9.index

0 directories,23 files
```

通常需要以不同的方式保存模型,以满足在生产开发中不同场景下深度学习模型的迁移和部署。下面用 saved_model 格式保存模型,用下述代码代替 model.save_weights('checkpoints/yolov3_train_{}'.format(epoch))。在代码编辑器中输入以下代码。

```
#model.save_weights('checkpoints/yolov3_train_{}'.format(epoch))
model.save('checkpoints/yolov3_train_{}'.format(epoch))
```

步骤 6:TensorBoard 可视化。

训练过程中测试集和验证集目标检测的损失全部记录在工程文件夹下的 ./logs/ 下,下面用 TensorBoard 对训练的整个过程进行可视化。在命令行中输入启动命令,可视化效果如图 6.23 和图 6.24 所示。

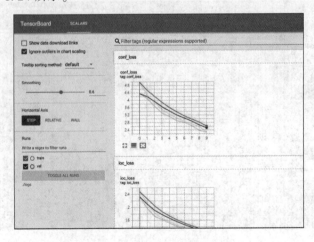

图 6.23 TensorBoard 可视化(一)

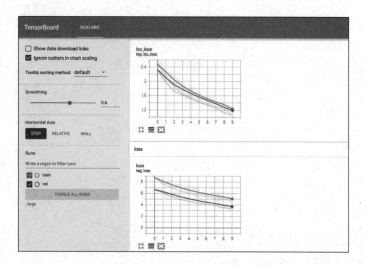

图 6.24　TensorBoard 可视化(二)

```
tensorboard --host 172.16.33.106 --port 8888 --logdir ./logs
```

步骤 7:测试模型。

使用训练好的模型进行测试,检查模型的推理效果。

首先在工程文件夹下新建 img 目录,在命令行中输入以下代码。

```
mkdir img
```

在 img 目录下放入一张命名为 test1.jpg 的检测图片。

新建文件 detect.py,在代码编辑器中输入如下代码。

```
import time
from absl import app,flags,logging
from absl.flags import FLAGS
import cv2
import numpy as np
import tensorflow as tf
from yolov3.models import (
    YoloV3,YoloV3Tiny
)
from yolov3.dataset import transform_images,load_tfrecord_dataset
from yolov3.utils import draw_outputs

flags.DEFINE_string('classes','./data/coco.names','path to classes file')
flags.DEFINE_string('weights','./checkpoints/yolov3',
                    'path to weights file')
flags.DEFINE_boolean('tiny',False,'yolov3 or yolov3-tiny')
flags.DEFINE_integer('size',416,'resize images to')
flags.DEFINE_string('image','./img/test1.jpg','path to input image')
flags.DEFINE_string('tfrecord',None,'tfrecord instead of image')
flags.DEFINE_string('output','./img/test1_result.jpg','path to output image')
flags.DEFINE_integer('num_classes',80,'number of classes in the model')
```

```python
def main(_argv):
    physical_devices = tf.config.experimental.list_physical_devices('GPU')
    for physical_device in physical_devices:
        tf.config.experimental.set_memory_growth(physical_device, True)
    if FLAGS.tiny:
        yolo = YoloV3Tiny(classes=FLAGS.num_classes)
    else:
        yolo = YoloV3(classes=FLAGS.num_classes)

    yolo.load_weights(FLAGS.weights).expect_partial()
    logging.info('weights loaded')

    class_names = [c.strip() for c in open(FLAGS.classes).readlines()]
    logging.info('classes loaded')

    if FLAGS.tfrecord:
        dataset = load_tfrecord_dataset(
            FLAGS.tfrecord, FLAGS.classes, FLAGS.size)
        dataset = dataset.shuffle(512)
        img_raw, _label = next(iter(dataset.take(1)))
    else:
        img_raw = tf.image.decode_image(
            open(FLAGS.image, 'rb').read(), channels=3)

    img = tf.expand_dims(img_raw, 0)
    img = transform_images(img, FLAGS.size)

    t1 = time.time()
    boxes, scores, classes, nums = yolo(img)
    t2 = time.time()
    logging.info('time: {}'.format(t2 - t1))

    logging.info('detections:')
    for i in range(nums[0]):
        logging.info('\t{}, {}, {}'.format(class_names[int(classes[0][i])],
                                           np.array(scores[0][i]), np.array(boxes[0][i])))

    img = cv2.cvtColor(img_raw.numpy(), cv2.COLOR_RGB2BGR)
    img = draw_outputs(img, (boxes, scores, classes, nums), class_names)
    cv2.imwrite(FLAGS.output, img)
    logging.info('output saved to: {}'.format(FLAGS.output))


if __name__ == '__main__':
    try:
        app.run(main)
    except SystemExit:
        pass
```

下面使用预训练模型 yolov3 测试,从数据资源平台下载 coco.names 标签对应文件,放在 data 目录下。

运行 detect.py 测试程序,在命令行中输入以下代码。

```
python detect.py
```

在 ./img/ 目录下生成 test1_result.jpg 的检测结果,如图 6.25 和图 6.26 所示。

图 6.25　待测图片

图 6.26　测试结果

单元任务 16　用 SSD 算法实现目标检测

步骤 1:创建工程和 ssd 虚拟环境。

使用 Anaconda 为 ssd 智能目标检测任务创建新的虚拟环境,在命令行中输入以下代码,运行结果如图 6.27 所示。

```
#创建工程文件
mkdir ~/projects/ssd

#创建名为ssd的虚拟环境
conda create -n ssd python=3.7
#输入 y 确认下载基础环境包

#进入ssd虚拟环境
conda activate ssd
```

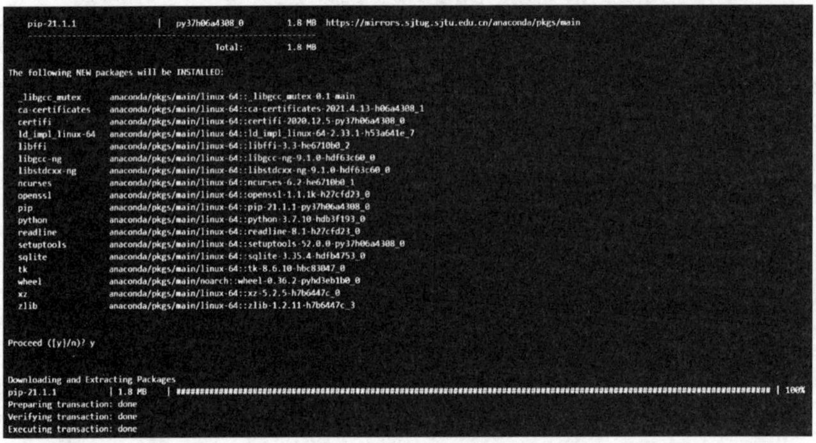

图 6.27　创建虚拟环境

安装 GPU 版 TensorFlow 和其 CUDA、cudnn 依赖,在命令行中输入安装命令,运行结果如图 6.28 所示。

```
conda install tensorflow-gpu=2.1
#输入 y 确认下载依赖包
```

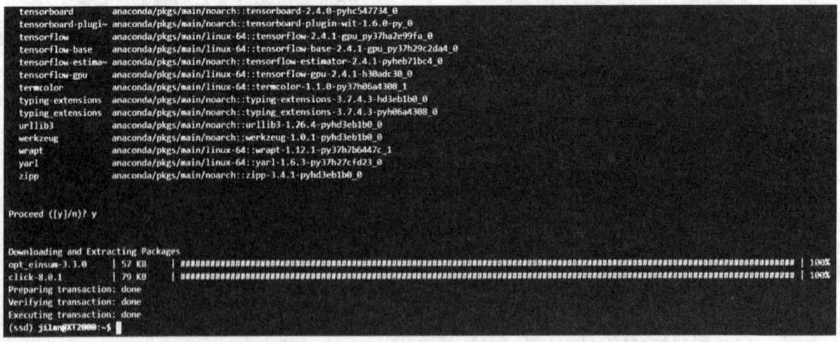

图 6.28　安装 GPU 版 TensorFlow 框架

使用 VS code 编辑器打开文件夹,建立一个 .py 文件,并为项目文件夹 ssd 指定 Python 解释器,如图 6.29 所示,ssd 路径为 ~/anaconda3/envs/bin/python。

重启 VS code 控制台终端,ssd 虚拟环境自动激活。

在工程文件夹 ssd 下新建 requirements.txt 文件,写入本项目需要的第三方依赖包,并

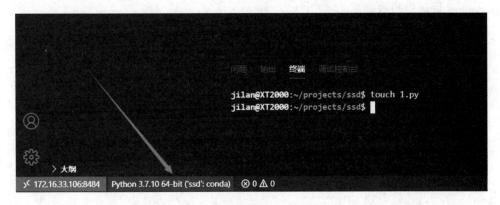

图 6.29　为项目指定 python 解释器

使用 pip 安装。

```
matplotlib
opencv-python
lxml
tqdm
Pillow
```

保存文件 requirements.txt 之后在命令行中输入以下代码。

```
pip install -r requirements.txt
```

安装完成后即可完成 ssd 算法目标检测的基本环境配置。

步骤 2：准备数据集。

在 ssd 工程文件夹下新建 datasets 数据集文件夹，首先从数据资源平台下载 VOC2007 和 VOC2012。

下载完成后在 datasets 目录下多出三份打包文件，分别是 VOC2007 训练集和验证集：VOCtrainval_06-Nov-2007.tar，VOC2007 测试集：VOCtest_06-Nov-2007.tar，VOC2012 训练集和验证集：VOCtrainval_11-May-2012.tar。

对 VOC2007 和 VOC2012 数据集包原始文件解包，在命令行中输入以下代码。

```
#继续在./datasets 文件夹下进行解包操作
#VOC2007 训练集验证集
tar -xvf VOCtrainval_06-Nov-2007.tar
#VOC2007 测试集
tar -xvf VOCtest_06-Nov-2007.tar
#VOC2012 训练集验证集
tar -xvf VOCtrainval_11-May-2012.tar
cd ..
```

解包完成后在 datasets 文件夹下可以得到 VOCdevkit 文件夹，该目录下有 VOC2007 和 VOC2012 的数据集目录。解包后完整的 VOC 数据集目录如下。其中 Annotations 目录、SegmentationClass 目录和 SegmentationObject 目录为标注文件，用于目标检测和图像分割等任务。ImageSets 目录为数据集标签索引目录。JPEGImages 为图像数据集目录，该目录下全部是格式为.jpg 的图片数据。在本任务中主要用到了 JPEGImages、ImageSets 和

Annotations 目录中的数据。

```
VOCdevkit/
├─VOC2007
│   ├─Annotations
│   ├─ImageSets
│   │   ├─Layout
│   │   ├─Main
│   │   └─Segmentation
│   ├─JPEGImages
│   ├─SegmentationClass
│   └─SegmentationObject
└─VOC2012
    ├─Annotations
    ├─ImageSets
    │   ├─Action
    │   ├─Layout
    │   ├─Main
    │   └─Segmentation
    ├─JPEGImages
    ├─SegmentationClass
    └─SegmentationObject
```

新建 ssd 模块文件夹,在命令行中输入以下代码。

```
mkdir ssd
```

新建__init__.py 文件,在命令行中输入以下代码。

```
cd ssd
touch __init__.py
cd ..
```

步骤 3:构建数据管道。

在 ssd 模块下新建 box_utils.py 文件,在命令行中输入以下代码。

```
cd ssd
touch box_utils.py
```

打开 box_utils.py 文件,在代码编辑器中输入以下代码。

```python
import tensorflow as tf

def compute_area(top_left,bot_right):
    """ Compute area given top_left and bottom_right coordinates
    Args:
        top_left: tensor (num_boxes,2)
        bot_right: tensor (num_boxes,2)
    Returns:
        area: tensor (num_boxes,)
    """
    #top_left: N x 2
    #bot_right: N x 2
    hw = tf.clip_by_value(bot_right - top_left,0.0,512.0)
    area = hw[...,0]* hw[...,1]

    return area
```

```python
def compute_iou(boxes_a,boxes_b):
    """ Compute overlap between boxes_a and boxes_b
    Args:
        boxes_a: tensor (num_boxes_a,4)
        boxes_b: tensor (num_boxes_b,4)
    Returns:
        overlap: tensor (num_boxes_a,num_boxes_b)
    """
    #boxes_a = > num_boxes_a,1,4
    boxes_a = tf.expand_dims(boxes_a,1)

    #boxes_b = > 1,num_boxes_b,4
    boxes_b = tf.expand_dims(boxes_b,0)
    top_left = tf.math.maximum(boxes_a[...,:2],boxes_b[...,:2])
    bot_right = tf.math.minimum(boxes_a[...,2:],boxes_b[...,2:])

    overlap_area = compute_area(top_left,bot_right)
    area_a = compute_area(boxes_a[...,:2],boxes_a[...,2:])
    area_b = compute_area(boxes_b[...,:2],boxes_b[...,2:])

    overlap = overlap_area / (area_a + area_b - overlap_area)

    return overlap
def compute_target(default_boxes,gt_boxes,gt_labels,iou_threshold = 0.5):
    """ Compute regression and classification targets
    Args:
        default_boxes: tensor (num_default,4)
                   of format (cx,cy,w,h)
        gt_boxes: tensor (num_gt,4)
                of format (xmin,ymin,xmax,ymax)
        gt_labels: tensor (num_gt,)
    Returns:
        gt_confs: classification targets,tensor (num_default,)
        gt_locs: regression targets,tensor (num_default,4)
    """
    #Convert default boxes to format (xmin,ymin,xmax,ymax)
    #in order to compute overlap with gt boxes
    transformed_default_boxes = transform_center_to_corner(default_boxes)
    iou = compute_iou(transformed_default_boxes,gt_boxes)

    best_gt_iou = tf.math.reduce_max(iou,1)
    best_gt_idx = tf.math.argmax(iou,1)

    best_default_iou = tf.math.reduce_max(iou,0)
    best_default_idx = tf.math.argmax(iou,0)

    best_gt_idx = tf.tensor_scatter_nd_update(
        best_gt_idx,
```

```
            tf.expand_dims(best_default_idx,1),
            tf.range(best_default_idx.shape[0],dtype=tf.int64))

    best_gt_iou=tf.tensor_scatter_nd_update(
        best_gt_iou,
        tf.expand_dims(best_default_idx,1),
        tf.ones_like(best_default_idx,dtype=tf.float32))

    gt_confs=tf.gather(gt_labels,best_gt_idx)
    gt_confs=tf.where(
        tf.less(best_gt_iou,iou_threshold),
        tf.zeros_like(gt_confs),
        gt_confs)
    gt_boxes=tf.gather(gt_boxes,best_gt_idx)
    gt_locs=encode(default_boxes,gt_boxes)

    return gt_confs,gt_locs

def encode(default_boxes,boxes,variance=[0.1,0.2]):
    """ Compute regression values
    Args:
        default_boxes: tensor (num_default,4)
                    of format (cx,cy,w,h)
        boxes: tensor (num_default,4)
            of format (xmin,ymin,xmax,ymax)
        variance: variance for center point and size
    Returns:
        locs: regression values,tensor (num_default,4)
    """
    #Convert boxes to (cx,cy,w,h) format
    transformed_boxes=transform_corner_to_center(boxes)

    locs=tf.concat([
        (transformed_boxes[...,:2]-default_boxes[:,:2]
        )/(default_boxes[:,2:]*variance[0]),
         tf.math.log(transformed_boxes[...,2:]/default_boxes[:,2:])/
variance[1]],axis=-1)

    return locs

def decode(default_boxes,locs,variance=[0.1,0.2]):
    """ Decode regression values back to coordinates
    Args:
        default_boxes: tensor (num_default,4)
                    of format (cx,cy,w,h)
        locs: tensor (batch_size,num_default,4)
            of format (cx,cy,w,h)
        variance: variance for center point and size
```

```python
    Returns:
        boxes: tensor (num_default, 4)
            of format (xmin, ymin, xmax, ymax)
    """
    locs = tf.concat([
        locs[..., :2] * variance[0] *
        default_boxes[:, 2:] + default_boxes[:, :2],
        tf.math.exp(locs[..., 2:] * variance[1]) * default_boxes[:, 2:]], axis=-1)

    boxes = transform_center_to_corner(locs)

    return boxes

def transform_corner_to_center(boxes):
    """ Transform boxes of format (xmin, ymin, xmax, ymax)
        to format (cx, cy, w, h)
    Args:
        boxes: tensor (num_boxes, 4)
            of format (xmin, ymin, xmax, ymax)
    Returns:
        boxes: tensor (num_boxes, 4)
            of format (cx, cy, w, h)
    """
    center_box = tf.concat([
        (boxes[..., :2] + boxes[..., 2:]) / 2,
        boxes[..., 2:] - boxes[..., :2]], axis=-1)

    return center_box

def transform_center_to_corner(boxes):
    """ Transform boxes of format (cx, cy, w, h)
        to format (xmin, ymin, xmax, ymax)
    Args:
        boxes: tensor (num_boxes, 4)
            of format (cx, cy, w, h)
    Returns:
        boxes: tensor (num_boxes, 4)
            of format (xmin, ymin, xmax, ymax)
    """
    corner_box = tf.concat([
        boxes[..., :2] - boxes[..., 2:] / 2,
        boxes[..., :2] + boxes[..., 2:] / 2], axis=-1)

    return corner_box
def compute_nms(boxes, scores, nms_threshold, limit=200):
    """ Perform Non Maximum Suppression algorithm
        to eliminate boxes with high overlap
```

```
    Args:
        boxes: tensor (num_boxes,4)
            of format (xmin,ymin,xmax,ymax)
        scores: tensor (num_boxes,)
        nms_threshold: NMS threshold
        limit: maximum number of boxes to keep

    Returns:
        idx: indices of kept boxes
    """
    if boxes.shape[0] == 0:
        return tf.constant([],dtype = tf.int32)
    selected = [0]
    idx = tf.argsort(scores,direction = 'DESCENDING')
    idx = idx[:limit]
    boxes = tf.gather(boxes,idx)

    iou = compute_iou(boxes,boxes)

    while True:
        row = iou[selected[-1]]
        next_indices = row <= nms_threshold
        #iou[:,~next_indices] = 1.0
        iou = tf.where(
            tf.expand_dims(tf.math.logical_not(next_indices),0),
            tf.ones_like(iou,dtype = tf.float32),iou)
        if not tf.math.reduce_any(next_indices):
            break

        selected.append(tf.argsort(
            tf.dtypes.cast(next_indices,tf.int32),direction = 'DESCENDING')[0].numpy())

    return tf.gather(idx,selected)
```

在 ssd 模块下新建 image_utils.py,在命令行中输入以下代码。

```
touch image_utils.py
```

打开 image_utils.py 文件,在代码编辑器中输入以下代码。

```
import os
from PIL import Image
import matplotlib.pyplot as plt
import matplotlib.patches as patches
import random
import numpy as np
import tensorflow as tf

from ssd.box_utils import compute_iou
```

```python
class ImageVisualizer(object):
    """ Class for visualizing image
    
    Attributes:
        idx_to_name: list to convert integer to string label
        class_colors: colors for drawing boxes and labels
        save_dir: directory to store images
    """
    
    def __init__(self, idx_to_name, class_colors=None, save_dir=None):
        self.idx_to_name = idx_to_name
        if class_colors is None or len(class_colors) != len(self.idx_to_name):
            self.class_colors = [[0, 255, 0]] * len(self.idx_to_name)
        else:
            self.class_colors = class_colors
        
        if save_dir is None:
            self.save_dir = './'
        else:
            self.save_dir = save_dir
        
        os.makedirs(self.save_dir, exist_ok=True)
    
    def save_image(self, img, boxes, labels, name):
        """ Method to draw boxes and labels
            then save to dir
        
        Args:
            img: numpy array (width, height, 3)
            boxes: numpy array (num_boxes, 4)
            labels: numpy array (num_boxes)
            name: name of image to be saved
        """
        plt.figure()
        fig, ax = plt.subplots(1)
        ax.imshow(img)
        save_path = os.path.join(self.save_dir, name)
        
        for i, box in enumerate(boxes):
            idx = labels[i] - 1
            cls_name = self.idx_to_name[idx]
            top_left = (box[0], box[1])
            bot_right = (box[2], box[3])
            ax.add_patch(patches.Rectangle(
                (box[0], box[1]),
                box[2] - box[0], box[3] - box[1],
                linewidth=2, edgecolor=(0., 1., 0.),
                facecolor="none"))
```

```python
            plt.text(
                box[0],
                box[1],
                s = cls_name,
                color = "white",
                verticalalignment = "top",
                bbox = {"color": (0.,1.,0.),"pad": 0},
            )
        plt.axis("off")
        #plt.gca().xaxis.set_major_locator(NullLocator())
        #plt.gca().yaxis.set_major_locator(NullLocator())
        plt.savefig(save_path,bbox_inches = "tight",pad_inches = 0.0)
        plt.close('all')

def generate_patch(boxes,threshold):
    """ Function to generate a random patch within the image
        If the patch overlaps any gt boxes at above the threshold,
        then the patch is picked,otherwise generate another patch

        Args:
            boxes: box tensor (num_boxes,4)
            threshold: iou threshold to decide whether to choose the patch

        Returns:
            patch: the picked patch
            ious: an array to store IOUs of the patch and all gt boxes
    """
    while True:
        patch_w = random.uniform(0.1,1)
        scale = random.uniform(0.5,2)
        patch_h = patch_w* scale
        patch_xmin = random.uniform(0,1 - patch_w)
        patch_ymin = random.uniform(0,1 - patch_h)
        patch_xmax = patch_xmin + patch_w
        patch_ymax = patch_ymin + patch_h
        patch = np.array(
            [[patch_xmin,patch_ymin,patch_xmax,patch_ymax]],
            dtype = np.float32)
        patch = np.clip(patch,0.0,1.0)
        ious = compute_iou(tf.constant(patch),boxes)
        if tf.math.reduce_any(ious > = threshold):
            break

    return patch[0],ious[0]
def random_patching(img,boxes,labels):
    """
        Args:
            img: the original PIL Image
```

```python
        boxes: gt boxes tensor (num_boxes,4)
        labels: gt labels tensor (num_boxes,)

    Returns:
        img: the cropped PIL Image
        boxes: selected gt boxes tensor (new_num_boxes,4)
        labels: selected gt labels tensor (new_num_boxes,)
    """
    threshold = np.random.choice(np.linspace(0.1,0.7,4))

    patch,ious = generate_patch(boxes,threshold)

    box_centers = (boxes[:,:2] + boxes[:,2:]) / 2
    keep_idx = (
        (ious > 0.3) &
        (box_centers[:,0] > patch[0]) &
        (box_centers[:,1] > patch[1]) &
        (box_centers[:,0] < patch[2]) &
        (box_centers[:,1] < patch[3])
    )

    if not tf.math.reduce_any(keep_idx):
        return img,boxes,labels

    img = img.crop(patch)

    boxes = boxes[keep_idx]
    patch_w = patch[2] - patch[0]
    patch_h = patch[3] - patch[1]
    boxes = tf.stack([
        (boxes[:,0] - patch[0]) / patch_w,
        (boxes[:,1] - patch[1]) / patch_h,
        (boxes[:,2] - patch[0]) / patch_w,
        (boxes[:,3] - patch[1]) / patch_h],axis=1)
    boxes = tf.clip_by_value(boxes,0.0,1.0)

    labels = labels[keep_idx]

    return img,boxes,labels
def horizontal_flip(img,boxes,labels):
    """
    Args:
        img: the original PIL Image
        boxes: gt boxes tensor (num_boxes,4)
        labels: gt labels tensor (num_boxes,)

    Returns:
        img: the horizontally flipped PIL Image
```

```
        boxes: horizontally flipped gt boxes tensor (num_boxes,4)
        labels: gt labels tensor (num_boxes,)
    """
    img = img.transpose(Image.FLIP_LEFT_RIGHT)
    boxes = tf.stack([
        1 - boxes[:,2],
        boxes[:,1],
        1 - boxes[:,0],
        boxes[:,3]],axis =1)

    return img,boxes,labels
```

在 ssd 模块下新建 voc_data.py 文件,在命令行中输入以下代码。

```
touch voc_data.py
```

打开 voc_data.py 文件,在代码编辑器中输入以下代码。

```
import tensorflow as tf
import os
import numpy as np
import xml.etree.ElementTree as ET
from PIL import Image
import random

from ssd.box_utils import compute_target
from ssd.image_utils import random_patching,horizontal_flip
from functools import partial

class VOCDataset():
    """ Class for VOC Dataset
    Attributes:
        root_dir: dataset root dir (ex: ./data/VOCdevkit)
        year: dataset's year (2007 or 2012)
        num_examples: number of examples to be used
                      (in case one wants to overfit small data)
    """

    def __init__(self,root_dir,year,default_boxes,
                 new_size,num_examples = -1,augmentation = None):
        super(VOCDataset,self).__init__()
        self.idx_to_name = [
            'aeroplane','bicycle','bird','boat',
            'bottle','bus','car','cat','chair',
            'cow','diningtable','dog','horse',
            'motorbike','person','pottedplant',
            'sheep','sofa','train','tvmonitor']
        self.name_to_idx = dict([(v,k)
                                 for k,v in enumerate(self.idx_to_name)])

        self.data_dir = os.path.join(root_dir,'VOC{}'.format(year))
```

```python
        self.image_dir = os.path.join(self.data_dir,'JPEGImages')
        self.anno_dir = os.path.join(self.data_dir,'Annotations')
        self.ids = list(map(lambda x: x[:-4],os.listdir(self.image_dir)))
        self.default_boxes = default_boxes
        self.new_size = new_size

        if num_examples != -1:
            self.ids = self.ids[:num_examples]

        self.train_ids = self.ids[:int(len(self.ids) * 0.75)]
        self.val_ids = self.ids[int(len(self.ids) * 0.75):]

        if augmentation == None:
            self.augmentation = ['original']
        else:
            self.augmentation = augmentation + ['original']
    def __len__(self):
        return len(self.ids)
    def _get_image(self,index):
        """ Method to read image from file
            then resize to (300,300)
            then subtract by ImageNet's mean
            then convert to Tensor

        Args:
            index: the index to get filename from self.ids

        Returns:
            img: tensor of shape (3,300,300)
        """
        filename = self.ids[index]
        img_path = os.path.join(self.image_dir,filename + '.jpg')
        img = Image.open(img_path)

        return img

    def _get_annotation(self,index,orig_shape):
        """ Method to read annotation from file
            Boxes are normalized to image size
            Integer labels are increased by 1

        Args:
            index: the index to get filename from self.ids
            orig_shape: image's original shape

        Returns:
            boxes: numpy array of shape (num_gt,4)
            labels: numpy array of shape (num_gt,)
        """
```

```python
            h,w = orig_shape
            filename = self.ids[index]
            anno_path = os.path.join(self.anno_dir,filename + '.xml')
            objects = ET.parse(anno_path).findall('object')
            boxes = []
            labels = []
            for obj in objects:
                name = obj.find('name').text.lower().strip()
                bndbox = obj.find('bndbox')
                xmin = (float(bndbox.find('xmin').text) - 1) / w
                ymin = (float(bndbox.find('ymin').text) - 1) / h
                xmax = (float(bndbox.find('xmax').text) - 1) / w
                ymax = (float(bndbox.find('ymax').text) - 1) / h
                boxes.append([xmin,ymin,xmax,ymax])

                labels.append(self.name_to_idx[name] + 1)

            return np.array(boxes,dtype = np.float32),np.array(labels,dtype = np.int64)

    def generate(self,subset = None):
        """ The __getitem__ method
            so that the object can be iterable

            Args:
                index: the index to get filename from self.ids

            Returns:
                img: tensor of shape (300,300,3)
                boxes: tensor of shape (num_gt,4)
                labels: tensor of shape (num_gt,)
        """
        if subset = = 'train':
            indices = self.train_ids
        elif subset = = 'val':
            indices = self.val_ids
        else:
            indices = self.ids
        for index in range(len(indices)):
            #img,orig_shape = self._get_image(index)
            filename = indices[index]
            img = self._get_image(index)
            w,h = img.size
            boxes,labels = self._get_annotation(index,(h,w))
            boxes = tf.constant(boxes,dtype = tf.float32)
            labels = tf.constant(labels,dtype = tf.int64)

            augmentation_method = np.random.choice(self.augmentation)
            if augmentation_method = = 'patch':
```

```
                    img,boxes,labels = random_patching(img,boxes,labels)
                elif augmentation_method = = 'flip':
                    img,boxes,labels = horizontal_flip(img,boxes,labels)

                img = np.array(img.resize(
                    (self.new_size,self.new_size)),dtype = np.float32)
                img = (img / 127.0) - 1.0
                img = tf.constant(img,dtype = tf.float32)

                gt_confs,gt_locs = compute_target(
                    self.default_boxes,boxes,labels)

                yield filename,img,gt_confs,gt_locs


def create_batch_generator(root_dir,year,default_boxes,
                           new_size,batch_size,num_batches,
                           mode,augmentation = None):
    num_examples = batch_size * num_batches if num_batches > 0 else -1
    voc = VOCDataset(root_dir,year,default_boxes,
                     new_size,num_examples,augmentation)

    info = {
        'idx_to_name': voc.idx_to_name,
        'name_to_idx': voc.name_to_idx,
        'length': len(voc),
        'image_dir': voc.image_dir,
        'anno_dir': voc.anno_dir
    }
    if mode = = 'train':
        train_gen = partial(voc.generate,subset = 'train')
        train_dataset = tf.data.Dataset.from_generator(
            train_gen,(tf.string,tf.float32,tf.int64,tf.float32))
        val_gen = partial(voc.generate,subset = 'val')
        val_dataset = tf.data.Dataset.from_generator(
            val_gen,(tf.string,tf.float32,tf.int64,tf.float32))

        train_dataset = train_dataset.shuffle(40).batch(batch_size)
        val_dataset = val_dataset.batch(batch_size)

        return train_dataset.take(num_batches),val_dataset.take(-1),info
    else:
        dataset = tf.data.Dataset.from_generator(
            voc.generate,(tf.string,tf.float32,tf.int64,tf.float32))
        dataset = dataset.batch(batch_size)
        return dataset.take(num_batches),info
```

步骤4:构建模型。

ssd 模型结构图如6.30所示。

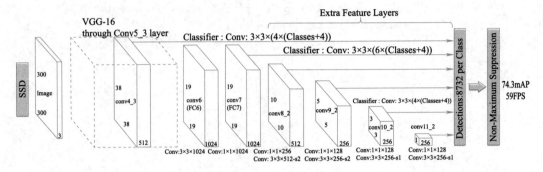

图 6.30　ssd 模型结构图

在 ssd 模块下新建 anchor.py 文件,在命令行中输入如下代码。

```
touch anchor.py
```

打开 anchor.py 文件,在代码编辑器中输入如下代码。

```
import itertools
import math
import tensorflow as tf
def generate_default_boxes(config):
    """ Generate default boxes for all feature maps
    Args:
        config: information of feature maps
            scales: boxes' size relative to image's size
            fm_sizes: sizes of feature maps
            ratios: box ratios used in each feature maps

    Returns:
        default_boxes: tensor of shape (num_default,4)
                    with format (cx,cy,w,h)
    """
    default_boxes = []
    scales = config['scales']
    fm_sizes = config['fm_sizes']
    ratios = config['ratios']

    for m,fm_size in enumerate(fm_sizes):
        for i,j in itertools.product(range(fm_size),repeat = 2):
            cx = (j + 0.5) / fm_size
            cy = (i + 0.5) / fm_size
            default_boxes.append([
                cx,
                cy,
                scales[m],
                scales[m]
            ])

            default_boxes.append([
```

```
                cx,
                cy,
                math.sqrt(scales[m] * scales[m + 1]),
                math.sqrt(scales[m] * scales[m + 1])
            ])

            for ratio in ratios[m]:
                r = math.sqrt(ratio)
                default_boxes.append([
                    cx,
                    cy,
                    scales[m] * r,
                    scales[m] / r
                ])
                default_boxes.append([
                    cx,
                    cy,
                    scales[m] / r,
                    scales[m] * r
                ])

    default_boxes = tf.constant(default_boxes)
    default_boxes = tf.clip_by_value(default_boxes, 0.0, 1.0)

    return default_boxes
```

在 ssd 模块下新建 layers.py 文件,在命令行中输入以下代码。

```
touch layers.py
```

打开 layers.py 文件,在代码编辑器中输入以下代码。

```
import tensorflow as tf
import tensorflow.keras.layers as layers
from tensorflow.keras import Sequential

def create_vgg16_layers():
    vgg16_conv4 = [
        layers.Conv2D(64, 3, padding = 'same', activation = 'relu'),
        layers.Conv2D(64, 3, padding = 'same', activation = 'relu'),
        layers.MaxPool2D(2, 2, padding = 'same'),

        layers.Conv2D(128, 3, padding = 'same', activation = 'relu'),
        layers.Conv2D(128, 3, padding = 'same', activation = 'relu'),
        layers.MaxPool2D(2, 2, padding = 'same'),

        layers.Conv2D(256, 3, padding = 'same', activation = 'relu'),
        layers.Conv2D(256, 3, padding = 'same', activation = 'relu'),
        layers.Conv2D(256, 3, padding = 'same', activation = 'relu'),
        layers.MaxPool2D(2, 2, padding = 'same'),
```

```python
        layers.Conv2D(512,3,padding = 'same',activation = 'relu'),
        layers.Conv2D(512,3,padding = 'same',activation = 'relu'),
        layers.Conv2D(512,3,padding = 'same',activation = 'relu'),
        layers.MaxPool2D(2,2,padding = 'same'),
        layers.Conv2D(512,3,padding = 'same',activation = 'relu'),
        layers.Conv2D(512,3,padding = 'same',activation = 'relu'),
        layers.Conv2D(512,3,padding = 'same',activation = 'relu'),
    ]

    x = layers.Input(shape = [None,None,3])
    out = x
    for layer in vgg16_conv4:
        out = layer(out)

    vgg16_conv4 = tf.keras.Model(x,out)

    vgg16_conv7 = [
        #Difference from original VGG16:
        #5th maxpool layer has kernel size = 3 and stride = 1
        layers.MaxPool2D(3,1,padding = 'same'),
        #atrous conv2d for 6th block
        layers.Conv2D(1024,3,padding = 'same',
                    dilation_rate = 6,activation = 'relu'),
        layers.Conv2D(1024,1,padding = 'same',activation = 'relu'),
    ]

    x = layers.Input(shape = [None,None,512])
    out = x
    for layer in vgg16_conv7:
        out = layer(out)

    vgg16_conv7 = tf.keras.Model(x,out)

    return vgg16_conv4,vgg16_conv7

def create_extra_layers():
    """ Create extra layers
        8th to 11th blocks
    """
    extra_layers = [
        #8th block output shape: B,512,10,10
        Sequential([
            layers.Conv2D(256,1,activation = 'relu'),
            layers.Conv2D(512,3,strides = 2,padding = 'same',
                        activation = 'relu'),
        ]),
```

```python
        #9th block output shape: B,256,5,5
        Sequential([
            layers.Conv2D(128,1,activation='relu'),
            layers.Conv2D(256,3,strides=2,padding='same',
                          activation='relu'),
        ]),
        #10th block output shape: B,256,3,3
        Sequential([
            layers.Conv2D(128,1,activation='relu'),
            layers.Conv2D(256,3,activation='relu'),
        ]),
        #11th block output shape: B,256,1,1
        Sequential([
            layers.Conv2D(128,1,activation='relu'),
            layers.Conv2D(256,3,activation='relu'),
        ]),
        #12th block output shape: B,256,1,1
        Sequential([
            layers.Conv2D(128,1,activation='relu'),
            layers.Conv2D(256,4,activation='relu'),
        ])
    ]

    return extra_layers

def create_conf_head_layers(num_classes):
    """ Create layers for classification
    """
    conf_head_layers = [
        layers.Conv2D(4*num_classes,kernel_size=3,
                      padding='same'),   #for 4th block
        layers.Conv2D(6*num_classes,kernel_size=3,
                      padding='same'),   #for 7th block
        layers.Conv2D(6*num_classes,kernel_size=3,
                      padding='same'),   #for 8th block
        layers.Conv2D(6*num_classes,kernel_size=3,
                      padding='same'),   #for 9th block
        layers.Conv2D(4*num_classes,kernel_size=3,
                      padding='same'),   #for 10th block
        layers.Conv2D(4*num_classes,kernel_size=3,
                      padding='same'),   #for 11th block
        layers.Conv2D(4*num_classes,kernel_size=1)   #for 12th block
    ]

    return conf_head_layers
def create_loc_head_layers():
    """ Create layers for regression
    """
```

```python
    loc_head_layers = [
        layers.Conv2D(4*4, kernel_size=3, padding='same'),
        layers.Conv2D(6*4, kernel_size=3, padding='same'),
        layers.Conv2D(6*4, kernel_size=3, padding='same'),
        layers.Conv2D(6*4, kernel_size=3, padding='same'),
        layers.Conv2D(4*4, kernel_size=3, padding='same'),
        layers.Conv2D(4*4, kernel_size=3, padding='same'),
        layers.Conv2D(4*4, kernel_size=1)
    ]

    return loc_head_layers
```

在 ssd 模块下新建 losses.py 文件，在命令行中输入以下代码。

```
touch losses.py
```

打开 losses.py 文件，在代码编辑器中输入以下代码。

```python
import tensorflow as tf

def hard_negative_mining(loss, gt_confs, neg_ratio):
    """
    Args:
        loss: list of classification losses of all default boxes (B, num_default)
        gt_confs: classification targets (B, num_default)
        neg_ratio: negative / positive ratio
    Returns:
        conf_loss: classification loss
        loc_loss: regression loss
    """
    #loss: B x N
    #gt_confs: B x N
    pos_idx = gt_confs > 0
    num_pos = tf.reduce_sum(tf.dtypes.cast(pos_idx, tf.int32), axis=1)
    num_neg = num_pos * neg_ratio

    rank = tf.argsort(loss, axis=1, direction='DESCENDING')
    rank = tf.argsort(rank, axis=1)
    neg_idx = rank < tf.expand_dims(num_neg, 1)

    return pos_idx, neg_idx
class SSDLosses(object):
    """ Class for SSD Losses
    Attributes:
        neg_ratio: negative / positive ratio
        num_classes: number of classes
    """

    def __init__(self, neg_ratio, num_classes):
        self.neg_ratio = neg_ratio
        self.num_classes = num_classes
```

```python
    def __call__(self,confs,locs,gt_confs,gt_locs):
        """ Compute losses for SSD
            regression loss: smooth L1
            classification loss: cross entropy
        Args:
            confs: outputs of classification heads (B,num_default,num_classes)
            locs: outputs of regression heads (B,num_default,4)
            gt_confs: classification targets (B,num_default)
            gt_locs: regression targets (B,num_default,4)
        Returns:
            conf_loss: classification loss
            loc_loss: regression loss
        """
        cross_entropy = tf.keras.losses.SparseCategoricalCrossentropy(
            from_logits = True,reduction = 'none')

        #compute classification losses
        #without reduction
        temp_loss = cross_entropy(
            gt_confs,confs)
        pos_idx,neg_idx = hard_negative_mining(
            temp_loss,gt_confs,self.neg_ratio)

        #classification loss will consist of positive and negative examples

        cross_entropy = tf.keras.losses.SparseCategoricalCrossentropy(
            from_logits = True,reduction = 'sum')
        smooth_l1_loss = tf.keras.losses.Huber(reduction = 'sum')
        conf_loss = cross_entropy(
            gt_confs[tf.math.logical_or(pos_idx,neg_idx)],
            confs[tf.math.logical_or(pos_idx,neg_idx)])

        #regression loss only consist of positive examples
        loc_loss = smooth_l1_loss(
            #tf.boolean_mask(gt_locs,pos_idx),
            #tf.boolean_mask(locs,pos_idx))
            gt_locs[pos_idx],
            locs[pos_idx])

        num_pos = tf.reduce_sum(tf.dtypes.cast(pos_idx,tf.float32))

        conf_loss = conf_loss / num_pos
        loc_loss = loc_loss / num_pos

        return conf_loss,loc_loss

def create_losses(neg_ratio,num_classes):
```

```
            criterion = SSDLosses(neg_ratio,num_classes)

        return criterion
```

在 ssd 模块下新建 network.py 文件,在命令行中输入以下代码。

```
touch network.py
```

打开 network.py 文件,在代码编辑器中输入以下代码。

```
from tensorflow.keras import Model
from tensorflow.keras.applications import VGG16
import tensorflow.keras.layers as layers
import tensorflow as tf
import numpy as np
import os
from ssd.layers import create_vgg16_layers,create_extra_layers,create_conf_head_layers,create_loc_head_layers

class SSD(Model):
    """ Class for SSD model
    Attributes:
        num_classes: number of classes
    """
    def __init__(self,num_classes,arch = 'ssd300'):
        super(SSD,self).__init__()
        self.num_classes = num_classes
        self.vgg16_conv4,self.vgg16_conv7 = create_vgg16_layers()
        self.batch_norm = layers.BatchNormalization(
            beta_initializer = 'glorot_uniform',
            gamma_initializer = 'glorot_uniform'
        )
        self.extra_layers = create_extra_layers()
        self.conf_head_layers = create_conf_head_layers(num_classes)
        self.loc_head_layers = create_loc_head_layers()

        if arch = = 'ssd300':
            self.extra_layers.pop(-1)
            self.conf_head_layers.pop(-2)
            self.loc_head_layers.pop(-2)

    def compute_heads(self,x,idx):
        """ Compute outputs of classification and regression heads
        Args:
            x: the input feature map
            idx: index of the head layer
        Returns:
            conf: output of the idx-th classification head
            loc: output of the idx-th regression head
        """
```

```python
        conf = self.conf_head_layers[idx](x)
        conf = tf.reshape(conf,[conf.shape[0],-1,self.num_classes])

        loc = self.loc_head_layers[idx](x)
        loc = tf.reshape(loc,[loc.shape[0],-1,4])

        return conf,loc

    def init_vgg16(self):
        """ Initialize the VGG16 layers from pretrained weights
            and the rest from scratch using xavier initializer
        """
        origin_vgg = VGG16(weights = 'imagenet')
        for i in range(len(self.vgg16_conv4.layers)):
            self.vgg16_conv4.get_layer(index = i).set_weights(
                origin_vgg.get_layer(index = i).get_weights())

        fc1_weights,fc1_biases = origin_vgg.get_layer(index = -3).get_weights()
        fc2_weights,fc2_biases = origin_vgg.get_layer(index = -2).get_weights()

        conv6_weights = np.random.choice(
            np.reshape(fc1_weights,(-1,)),(3,3,512,1024))
        conv6_biases = np.random.choice(
            fc1_biases,(1024,))

        conv7_weights = np.random.choice(
            np.reshape(fc2_weights,(-1,)),(1,1,1024,1024))
        conv7_biases = np.random.choice(
            fc2_biases,(1024,))

        self.vgg16_conv7.get_layer(index = 2).set_weights(
            [conv6_weights,conv6_biases])
        self.vgg16_conv7.get_layer(index = 3).set_weights(
            [conv7_weights,conv7_biases])

    def call(self,x):
        """ The forward pass
        Args:
            x: the input image
        Returns:
            confs: list of outputs of all classification heads
            locs: list of outputs of all regression heads
        """
        confs = []
        locs = []
        head_idx = 0
        for i in range(len(self.vgg16_conv4.layers)):
            x = self.vgg16_conv4.get_layer(index = i)(x)
```

```python
            if i == len(self.vgg16_conv4.layers) - 5:
                conf, loc = self.compute_heads(self.batch_norm(x), head_idx)
                confs.append(conf)
                locs.append(loc)
                head_idx += 1

        x = self.vgg16_conv7(x)

        conf, loc = self.compute_heads(x, head_idx)

        confs.append(conf)
        locs.append(loc)
        head_idx += 1

        for layer in self.extra_layers:
            x = layer(x)
            conf, loc = self.compute_heads(x, head_idx)
            confs.append(conf)
            locs.append(loc)
            head_idx += 1

        confs = tf.concat(confs, axis=1)
        locs = tf.concat(locs, axis=1)

        return confs, locs

def create_ssd(num_classes, arch, pretrained_type,
               checkpoint_dir=None,
               checkpoint_path=None):
    """ Create SSD model and load pretrained weights
    Args:
        num_classes: number of classes
        pretrained_type: type of pretrained weights, can be either 'VGG16' or 'ssd'
        weight_path: path to pretrained weights
    Returns:
        net: the SSD model
    """
    net = SSD(num_classes, arch)
    net(tf.random.normal((1, 512, 512, 3)))
    if pretrained_type == 'base':
        net.init_vgg16()
    elif pretrained_type == 'latest':
        try:
            paths = [os.path.join(checkpoint_dir, path)
                     for path in os.listdir(checkpoint_dir)]
            latest = sorted(paths, key=os.path.getmtime)[-1]
            net.load_weights(latest)
```

```
            except AttributeError as e:
                print('Please make sure there is at least one checkpoint at {}'
.format(checkpoint_dir))
                print('The model will be loaded from base weights.')
                net.init_vgg16()
            except ValueError as e:
                raise ValueError(
                    'Please check the following \n1. / Is the path correct: {}? \n2. /
Is the model architecture correct: {}? '.format(latest,arch))
            except Exception as e:
                print(e)
                raise ValueError('Please check if checkpoint_dir is specified')
        elif pretrained_type == 'specified':
            if not os.path.isfile(checkpoint_path):
                raise ValueError(
                    'Not a valid checkpoint file: {}'.format(checkpoint_path))

            try:
                net.load_weights(checkpoint_path)
            except Exception as e:
                raise ValueError(
                    'Please check the following \n1. / Is the path correct: {}? \n2. /
Is the model architecture correct: {}? '.format(
                        checkpoint_path,arch))
        else:
            raise ValueError('Unknown pretrained type: {}'.format(pretrained_type))
        return net
```

在 ssd 模块下新建 voc_eval.py 文件,在命令行中输入以下代码。

```
touch voc_eval.py
```

打开 voc_eval.py 文件,在代码编辑器中输入以下代码。

```
import os
import numpy as np
import xml.etree.ElementTree as ET
import argparse

parser = argparse.ArgumentParser()
parser.add_argument('--data-dir',default='../dataset')
parser.add_argument('--data-year',default='2007')
parser.add_argument('--detect-dir',default='./outputs/detects')
parser.add_argument('--use-07-metric',type=bool,default=False)
args = parser.parse_args()

def get_annotation(anno_file):
    tree = ET.parse(anno_file)
    objects = []
    for obj in tree.findall('object'):
        obj_struct = {}
```

```python
            obj_struct['name'] = obj.find('name').text
            obj_struct['pose'] = obj.find('pose').text
            obj_struct['truncated'] = int(obj.find('truncated').text)
            obj_struct['difficult'] = int(obj.find('difficult').text)
            bbox = obj.find('bndbox')
            obj_struct['bbox'] = [int(bbox.find('xmin').text),
                                  int(bbox.find('ymin').text),
                                  int(bbox.find('xmax').text),
                                  int(bbox.find('ymax').text)]
            objects.append(obj_struct)

    return objects
def compute_ap(rec,prec,ap,use_07_metric = False):
    if use_07_metric:
        ap = 0.0
        for t in np.arange(0.0,1.1,0.1):
            if np.sum(rec > = t) = = 0:
                p = 0
            else:
                p = np.max(prec[rec > = t])
            ap = ap + p / 11.0
    else:
        mrec = np.concatenate(([0.0],rec,[1.0]))
        mprec = np.concatenate(([0.0],prec,[0.0]))

        for i in range(mprec.size - 1,0,-1):
            mprec[i - 1] = np.maximum(mprec[i - 1],mprec[i])

        i = np.where(mrec[1:] ! = mrec[:-1])[0]

        ap = np.sum((mrec[i + 1] - mrec[i]) * mprec[i + 1])

    return ap

def voc_eval(det_path,anno_path,cls_name,iou_thresh = 0.5,use_07_metric = False):
    det_file = det_path.format(cls_name)
    with open(det_file,'r') as f:
        lines = f.readlines()

    lines = [x.strip().split(' ') for x in lines]
    image_ids = [x[0] for x in lines]
    confs = np.array([float(x[1]) for x in lines])
    boxes = np.array([[float(z) for z in x[2:]] for x in lines])

    gts = {}
    cls_gts = {}
    npos = 0
    for image_id in image_ids:
```

```python
            if image_id in cls_gts.keys():
                continue
            gts[image_id] = get_annotation(anno_path.format(image_id))
        R = [obj for obj in gts[image_id] if obj['name'] == cls_name]
        gt_boxes = np.array([x['bbox'] for x in R])
        difficult = np.array([x['difficult'] for x in R]).astype(np.bool)
        det = [False] * len(R)
        npos = npos + sum(~difficult)
        cls_gts[image_id] = {
            'gt_boxes': gt_boxes,
            'difficult': difficult,
            'det': det
        }

sorted_ids = np.argsort(-confs)
sorted_scores = np.sort(-confs)
boxes = boxes[sorted_ids, :]
image_ids = [image_ids[x] for x in sorted_ids]

nd = len(image_ids)
tp = np.zeros(nd)
fp = np.zeros(nd)
for d in range(nd):
    R = cls_gts[image_ids[d]]
    box = boxes[d, :].astype(float)
    iou_max = -np.inf
    gt_box = R['gt_boxes'].astype(float)

    if gt_box.size > 0:
        ixmin = np.maximum(gt_box[:,0], box[0])
        ixmax = np.minimum(gt_box[:,2], box[2])
        iymin = np.maximum(gt_box[:,1], box[1])
        iymax = np.minimum(gt_box[:,3], box[3])
        iw = np.maximum(ixmax - ixmin + 1.0, 0.0)
        ih = np.maximum(iymax - iymin + 1.0, 0.0)
        inters = iw * ih

        uni = ((box[2] - box[0] + 1.0) * (box[3] - box[1] + 1.0) +
               (gt_box[:,2] - gt_box[:,0] + 1.0) *
               (gt_box[:,3] - gt_box[:,1] + 1.0) - inters)
        ious = inters / uni
        iou_max = np.max(ious)
        jmax = np.argmax(ious)

    if iou_max > iou_thresh:
        if not R['difficult'][jmax]:
            if not R['det'][jmax]:
                tp[d] = 1.0
```

```python
                        R['det'][jmax] = 1
                else:
                    fp[d] = 1.0
        else:
            fp[d] = 1.0

    fp = np.cumsum(fp)
    tp = np.cumsum(tp)
    recall = tp / float(npos)
    precision = tp / np.maximum(tp + fp, np.finfo(np.float64).eps)

    ap = compute_ap(recall, precision, use_07_metric)

    return recall, precision, ap

if __name__ == '__main__':
    aps = {
        'aeroplane': 0.0,
        'bicycle': 0.0,
        'bird': 0.0,
        'boat': 0.0,
        'bottle': 0.0,
        'bus': 0.0,
        'car': 0.0,
        'cat': 0.0,
        'chair': 0.0,
        'cow': 0.0,
        'diningtable': 0.0,
        'dog': 0.0,
        'horse': 0.0,
        'motorbike': 0.0,
        'person': 0.0,
        'pottedplant': 0.0,
        'sheep': 0.0,
        'sofa': 0.0,
        'train': 0.0,
        'tvmonitor': 0.0,
        'mAP': []
    }
    for cls_name in aps.keys():
        det_path = os.path.join(args.detect_dir, '{}.txt')
        anno_path = os.path.join(
            args.data_dir, 'VOC{}'.format(args.data_year), 'Annotations', '{}.xml')
        if os.path.exists(det_path.format(cls_name)):
            recall, precision, ap = voc_eval(
                det_path,
                anno_path,
```

```
                cls_name,
                use_07_metric = args.use_07_metric)
        aps[cls_name] = ap
        aps['mAP'].append(ap)

    aps['mAP'] = np.mean(aps['mAP'])
    for key,value in aps.items():
        print('{}: {}'.format(key,value))
```

开始构建 ssd 全局模块,返回到工程文件目录下,新建 setup.py 文件构建全局模块,在命令行中输入以下代码。

```
cd ..
touch setup.py
```

在 setup.py 中使用 setuptool 为前面写好的 ssd 目录构建全局模块,模块名为 ssd,在代码编辑器中输入以下代码。

```
from setuptools import setup

setup(name = 'ssd',
      version = '1.0',
      packages = ['ssd'])
```

运行 setup.py 文件为 yolov3 构建模块并安装到全局虚拟环境 yolov3 中,在命令行中输入以下代码。

```
python setup.py develop
```

测试前面 ssd 模块是否成功安装到全局虚拟环境中,在命令行中输入以下代码。

```
python
#进入交互环境
Python 3.7.10 (default,Feb 26 2021,18:47:35)
[GCC 7.3.0] :: Anaconda,Inc. on linux
Type "help","copyright","credits" or "license" for more information.
>>>import ssd
>>>
#可以成功导入 ssd 模块则表示该模块成功安装到全局虚拟环境中
```

步骤 5:训练模型。

在工程文件夹下新建 train.py 文件,在代码编辑器中输入以下代码。

```
import argparse
import tensorflow as tf
import os
import sys
import time
from tensorflow.keras.optimizers.schedules import PiecewiseConstantDecay
from ssd.voc_data import create_batch_generator
from ssd.anchor import generate_default_boxes
from ssd.network import create_ssd
from ssd.losses import create_losses
```

```python
cfg = {'SSD300':
    {'ratios': [[2],[2,3],[2,3],[2,3],[2],[2]],
    'scales': [0.1,0.2,0.375,0.55,0.725,0.9,1.075],
    'fm_sizes': [38,19,10,5,3,1],
    'image_size': 300
    },
    'SSD512':
    {'ratios': [[2],[2,3],[2,3],[2,3],[2],[2],[2]],
    'scales': [0.07,0.15,0.3,0.45,0.6,0.75,0.9,1.05],
    'fm_sizes': [64,32,16,8,6,4,1],
    'image_size': 512
}}
parser = argparse.ArgumentParser()
parser.add_argument('--data_dir',default = './datasets/VOCdevkit')
parser.add_argument('--data-year',default = '2007')
parser.add_argument('--arch',default = 'ssd300')
parser.add_argument('--batch-size',default = 8,type = int)
parser.add_argument('--num-batches',default = -1,type = int)
parser.add_argument('--neg-ratio',default = 3,type = int)
parser.add_argument('--initial-lr',default = 1e-3,type = float)
parser.add_argument('--momentum',default = 0.9,type = float)
parser.add_argument('--weight-decay',default = 5e-4,type = float)
parser.add_argument('--num-epochs',default = 10,type = int)
parser.add_argument('--checkpoint-dir',default = 'checkpoints')
parser.add_argument('--pretrained-type',default = 'base')
parser.add_argument('--gpu-id',default = '0')

args = parser.parse_args()

os.environ['CUDA_VISIBLE_DEVICES'] = args.gpu_id

NUM_CLASSES = 21

@tf.function
def train_step(imgs,gt_confs,gt_locs,ssd,criterion,optimizer):
    with tf.GradientTape() as tape:
        confs,locs = ssd(imgs)

        conf_loss,loc_loss = criterion(confs,locs,gt_confs,gt_locs)

        loss = conf_loss + loc_loss
        l2_loss = [tf.nn.l2_loss(t) for t in ssd.trainable_variables]
        l2_loss = args.weight_decay* tf.math.reduce_sum(l2_loss)
        loss += l2_loss
```

```python
        gradients = tape.gradient(loss,ssd.trainable_variables)
        optimizer.apply_gradients(zip(gradients,ssd.trainable_variables))

    return loss,conf_loss,loc_loss,l2_loss
if __name__ == '__main__':
    physical_devices = tf.config.experimental.list_physical_devices('GPU')
    for physical_device in physical_devices:
        tf.config.experimental.set_memory_growth(physical_device,True)

    os.makedirs(args.checkpoint_dir,exist_ok = True)

    try:
        config = cfg[args.arch.upper()]
    except AttributeError:
        raise ValueError('Unknown architecture: {}'.format(args.arch))

    default_boxes = generate_default_boxes(config)

    batch_generator,val_generator,info = create_batch_generator(
        args.data_dir,args.data_year,default_boxes,
        config['image_size'],
        args.batch_size,args.num_batches,
        mode = 'train',augmentation = ['flip'])

    try:
        ssd = create_ssd(NUM_CLASSES,args.arch,
                    args.pretrained_type,
                    checkpoint_dir = args.checkpoint_dir)
    except Exception as e:
        print(e)
        print('The program is exiting...')
        sys.exit()

    criterion = create_losses(args.neg_ratio,NUM_CLASSES)

    steps_per_epoch = info['length'] //args.batch_size

    lr_fn = PiecewiseConstantDecay(
        boundaries = [int(steps_per_epoch* args.num_epochs * 2/3),
                    int(steps_per_epoch* args.num_epochs * 5/6)],
        values = [args.initial_lr,args.initial_lr* 0.1,args.initial_lr * 0.01])
    optimizer = tf.keras.optimizers.SGD(
        learning_rate = lr_fn,
        momentum = args.momentum)

    train_log_dir = 'logs/train'
    val_log_dir = 'logs/val'
    train_summary_writer = tf.summary.create_file_writer(train_log_dir)
```

```
            val_summary_writer = tf.summary.create_file_writer(val_log_dir)

        for epoch in range(args.num_epochs):
            avg_loss = 0.0
            avg_conf_loss = 0.0
            avg_loc_loss = 0.0
            start = time.time()
            for i,(_,imgs,gt_confs,gt_locs) in enumerate(batch_generator):
                loss,conf_loss,loc_loss,l2_loss = train_step(
                    imgs,gt_confs,gt_locs,ssd,criterion,optimizer)
                avg_loss = (avg_loss* i + loss.numpy())/(i + 1)
                avg_conf_loss = (avg_conf_loss* i + conf_loss.numpy())/(i + 1)
                avg_loc_loss = (avg_loc_loss* i + loc_loss.numpy())/(i + 1)
                if (i + 1)% 50 == 0:
                    print('Epoch: {} Batch {} Time: {:.2}s | Loss: {:.4f} Confidence_loss: {:.4f} Localization_loss: {:.4f}'.format(epoch + 1,i + 1,time.time() - start, avg_loss,avg_conf_loss,avg_loc_loss))

            avg_val_loss = 0.0
            avg_val_conf_loss = 0.0
            avg_val_loc_loss = 0.0
            for i,(_,imgs,gt_confs,gt_locs) in enumerate(val_generator):
                val_confs,val_locs = ssd(imgs)
                val_conf_loss,val_loc_loss = criterion(
                    val_confs,val_locs,gt_confs,gt_locs)
                val_loss = val_conf_loss + val_loc_loss
                avg_val_loss = (avg_val_loss * i + val_loss.numpy())/(i + 1)
                avg_val_conf_loss = (avg_val_conf_loss* i + val_conf_loss.numpy())/(i + 1)
                avg_val_loc_loss = (avg_val_loc_loss* i + val_loc_loss.numpy())/(i + 1)
            with train_summary_writer.as_default():
                tf.summary.scalar('loss',avg_loss,step = epoch)
                tf.summary.scalar('conf_loss',avg_conf_loss,step = epoch)
                tf.summary.scalar('loc_loss',avg_loc_loss,step = epoch)

            with val_summary_writer.as_default():
                tf.summary.scalar('loss',avg_val_loss,step = epoch)
                tf.summary.scalar('conf_loss',avg_val_conf_loss,step = epoch)
                tf.summary.scalar('loc_loss',avg_val_loc_loss,step = epoch)

            if (epoch + 1) %10 == 0:
                ssd.save_weights(os.path.join(args.checkpoint_dir,'ssd_epoch_{}.h5'.format(epoch + 1)))
                #ssd.save('./saved_model/saved_model_epoch{}'.format(epoch + 1))
```

在命令行中输入运行命令,模型训练过程如图6.31所示。

```
python train.py
```

```
Epoch: 9 Batch 50 Time: 2.5e+01s | Loss: 5.1108 Confidence_loss: 2.5903 Localization_loss: 1.2767
Epoch: 9 Batch 100 Time: 4.9e+01s | Loss: 5.1523 Confidence_loss: 2.6199 Localization_loss: 1.2889
Epoch: 9 Batch 150 Time: 7.3e+01s | Loss: 5.1500 Confidence_loss: 2.6313 Localization_loss: 1.2752
Epoch: 9 Batch 200 Time: 9.7e+01s | Loss: 5.1630 Confidence_loss: 2.6421 Localization_loss: 1.2774
Epoch: 9 Batch 250 Time: 1.2e+02s | Loss: 5.1322 Confidence_loss: 2.6287 Localization_loss: 1.2601
Epoch: 9 Batch 300 Time: 1.5e+02s | Loss: 5.1179 Confidence_loss: 2.6200 Localization_loss: 1.2546
Epoch: 9 Batch 350 Time: 1.7e+02s | Loss: 5.1067 Confidence_loss: 2.6159 Localization_loss: 1.2475
Epoch: 9 Batch 400 Time: 1.9e+02s | Loss: 5.1134 Confidence_loss: 2.6190 Localization_loss: 1.2512
Epoch: 9 Batch 450 Time: 2.2e+02s | Loss: 5.1070 Confidence_loss: 2.6140 Localization_loss: 1.2499
Epoch: 9 Batch 500 Time: 2.4e+02s | Loss: 5.1031 Confidence_loss: 2.6126 Localization_loss: 1.2474
Epoch: 9 Batch 550 Time: 2.7e+02s | Loss: 5.0948 Confidence_loss: 2.6054 Localization_loss: 1.2463
Epoch: 9 Batch 600 Time: 2.9e+02s | Loss: 5.0952 Confidence_loss: 2.6057 Localization_loss: 1.2465
Epoch: 9 Batch 650 Time: 3.2e+02s | Loss: 5.0965 Confidence_loss: 2.6080 Localization_loss: 1.2456
Epoch: 9 Batch 700 Time: 3.4e+02s | Loss: 5.1013 Confidence_loss: 2.6136 Localization_loss: 1.2448
Epoch: 9 Batch 750 Time: 3.6e+02s | Loss: 5.0981 Confidence_loss: 2.6113 Localization_loss: 1.2439
Epoch: 9 Batch 800 Time: 3.9e+02s | Loss: 5.0920 Confidence_loss: 2.6060 Localization_loss: 1.2433
Epoch: 9 Batch 850 Time: 4.1e+02s | Loss: 5.0914 Confidence_loss: 2.6054 Localization_loss: 1.2432
Epoch: 9 Batch 900 Time: 4.4e+02s | Loss: 5.0838 Confidence_loss: 2.6004 Localization_loss: 1.2408
Epoch: 10 Batch 50 Time: 2.5e+01s | Loss: 4.7310 Confidence_loss: 2.3441 Localization_loss: 1.1452
Epoch: 10 Batch 100 Time: 4.9e+01s | Loss: 4.7411 Confidence_loss: 2.3545 Localization_loss: 1.1448
Epoch: 10 Batch 150 Time: 7.3e+01s | Loss: 4.7387 Confidence_loss: 2.3657 Localization_loss: 1.1313
Epoch: 10 Batch 200 Time: 9.8e+01s | Loss: 4.7181 Confidence_loss: 2.3569 Localization_loss: 1.1195
Epoch: 10 Batch 250 Time: 1.2e+02s | Loss: 4.6822 Confidence_loss: 2.3445 Localization_loss: 1.0960
Epoch: 10 Batch 300 Time: 1.5e+02s | Loss: 4.6531 Confidence_loss: 2.3259 Localization_loss: 1.0855
Epoch: 10 Batch 350 Time: 1.7e+02s | Loss: 4.6403 Confidence_loss: 2.3184 Localization_loss: 1.0802
Epoch: 10 Batch 400 Time: 1.9e+02s | Loss: 4.6422 Confidence_loss: 2.3215 Localization_loss: 1.0790
Epoch: 10 Batch 450 Time: 2.2e+02s | Loss: 4.6271 Confidence_loss: 2.3137 Localization_loss: 1.0717
Epoch: 10 Batch 500 Time: 2.4e+02s | Loss: 4.6167 Confidence_loss: 2.3070 Localization_loss: 1.0681
Epoch: 10 Batch 550 Time: 2.7e+02s | Loss: 4.6095 Confidence_loss: 2.3022 Localization_loss: 1.0658
Epoch: 10 Batch 600 Time: 2.9e+02s | Loss: 4.5999 Confidence_loss: 2.2967 Localization_loss: 1.0616
Epoch: 10 Batch 650 Time: 3.2e+02s | Loss: 4.6001 Confidence_loss: 2.2967 Localization_loss: 1.0619
Epoch: 10 Batch 700 Time: 3.4e+02s | Loss: 4.5906 Confidence_loss: 2.2925 Localization_loss: 1.0565
Epoch: 10 Batch 750 Time: 3.6e+02s | Loss: 4.5848 Confidence_loss: 2.2896 Localization_loss: 1.0536
Epoch: 10 Batch 800 Time: 3.9e+02s | Loss: 4.5739 Confidence_loss: 2.2814 Localization_loss: 1.0509
Epoch: 10 Batch 850 Time: 4.1e+02s | Loss: 4.5702 Confidence_loss: 2.2784 Localization_loss: 1.0502
Epoch: 10 Batch 900 Time: 4.4e+02s | Loss: 4.5731 Confidence_loss: 2.2808 Localization_loss: 1.0508
```

图 6.31　训练模型

步骤 6：TensorBoard 可视化。

训练过程中测试集和验证集目标检测的损失全部记录在工程文件夹下的 ./logs/ 下，下面用 TensorBoard 对训练的整个过程进行可视化。在命令行中输入启动命令，可视化如图 6.32 和图 6.33 所示。

```
tensorboard --host 172.16.33.106 --port 8888 --logdir ./logs
```

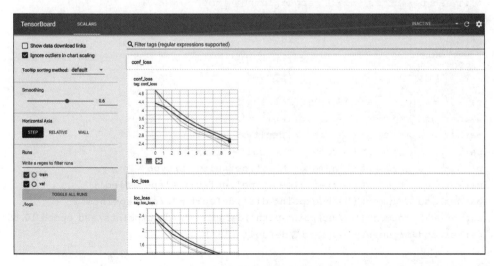

图 6.32　TensorBoard 可视化（一）

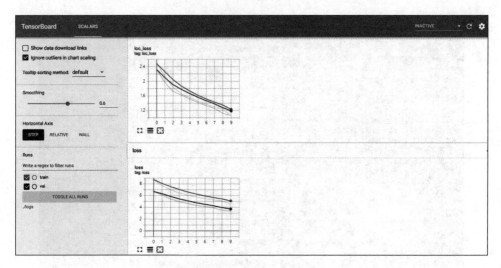

图 6.33 TensorBoard 可视化（二）

步骤 7：测试模型。

在工程目录下新建 img 目录和 detect.py 文件，在命令行中输入如下代码。

```
mkdir img
touch detect.py
```

打开 detect.py 文件，在代码编辑器中输入如下代码。

```
import argparse
import tensorflow as tf
import os
import sys
import numpy as np
from tqdm import tqdm
from ssd.anchor import generate_default_boxes
from ssd.box_utils import decode, compute_nms
from ssd.image_utils import ImageVisualizer
from ssd.network import create_ssd
from PIL import Image


parser = argparse.ArgumentParser()
parser.add_argument('--img', default = './img/test2.jpg')
parser.add_argument('--arch', default = 'ssd300')
parser.add_argument('--result', default = './img')
parser.add_argument('--num-examples', default = -1, type = int)
parser.add_argument('--pretrained-type', default = 'specified')
parser.add_argument('--checkpoint-dir', default = './checkpoints')
parser.add_argument('--checkpoint-path', default = './checkpoints/ssd_epoch_10.h5')
parser.add_argument('--gpu-id', default = '0')
```

```python
args = parser.parse_args()

os.environ['CUDA_VISIBLE_DEVICES'] = args.gpu_id

NUM_CLASSES = 21
BATCH_SIZE = 1
cfg = {'SSD300':
    {'ratios': [[2],[2,3],[2,3],[2,3],[2],[2]],
     'scales': [0.1,0.2,0.375,0.55,0.725,0.9,1.075],
     'fm_sizes': [38,19,10,5,3,1],
     'image_size': 300
    },
    'SSD512':
    {'ratios': [[2],[2,3],[2,3],[2,3],[2],[2],[2]],
     'scales': [0.07,0.15,0.3,0.45,0.6,0.75,0.9,1.05],
     'fm_sizes': [64,32,16,8,6,4,1],
     'image_size': 512
    }}
idx_to_name = [
        'aeroplane','bicycle','bird','boat',
        'bottle','bus','car','cat','chair',
        'cow','diningtable','dog','horse',
        'motorbike','person','pottedplant',
        'sheep','sofa','train','tvmonitor']

def get_img(img,default_size):
    img = Image.open(img)
    img = np.array(img.resize(default_size),dtype = np.float32)
    img = (img / 127.0) - 1.0
    img = tf.constant(img,dtype = tf.float32)
    img = tf.data.Dataset.from_tensors(img).batch(BATCH_SIZE).take(BATCH_SIZE)
    return img


def predict(imgs,default_boxes):
    confs,locs = ssd(imgs)

    confs = tf.squeeze(confs,0)
    locs = tf.squeeze(locs,0)

    confs = tf.math.softmax(confs,axis = -1)
    classes = tf.math.argmax(confs,axis = -1)
    scores = tf.math.reduce_max(confs,axis = -1)

    boxes = decode(default_boxes,locs)

    out_boxes = []
    out_labels = []
```

```python
        out_scores = []
        for c in range(1,NUM_CLASSES):
            cls_scores = confs[:,c]

            score_idx = cls_scores > 0.6
            #cls_boxes = tf.boolean_mask(boxes,score_idx)
            #cls_scores = tf.boolean_mask(cls_scores,score_idx)
            cls_boxes = boxes[score_idx]
            cls_scores = cls_scores[score_idx]

            nms_idx = compute_nms(cls_boxes,cls_scores,0.45,200)
            cls_boxes = tf.gather(cls_boxes,nms_idx)
            cls_scores = tf.gather(cls_scores,nms_idx)
            cls_labels = [c]* cls_boxes.shape[0]

            out_boxes.append(cls_boxes)
            out_labels.extend(cls_labels)
            out_scores.append(cls_scores)

        out_boxes = tf.concat(out_boxes,axis = 0)
        out_scores = tf.concat(out_scores,axis = 0)

        boxes = tf.clip_by_value(out_boxes,0.0,1.0).numpy()
        classes = np.array(out_labels)
        scores = out_scores.numpy()

        return boxes,classes,scores

    if __name__ == '__main__':
        try:
            config = cfg[args.arch.upper()]
        except AttributeError:
            raise ValueError('Unknown architecture: {}'.format(args.arch))

        default_boxes = generate_default_boxes(config)
        default_size = (cfg[args.arch.upper()]['image_size'],cfg[args.arch.upper()]['image_size'])
        try:
            ssd = create_ssd(NUM_CLASSES,args.arch,
                            args.pretrained_type,
                            args.checkpoint_dir,
                            args.checkpoint_path)
        except Exception as e:
            print(e)
            print('The program is exiting...')
            sys.exit()
```

```
visualizer = ImageVisualizer(idx_to_name = idx_to_name,save_dir = args.result)
img = get_img(args.img,default_size)
for i,img in enumerate(img):
    boxes,classes,scores = predict(img,default_boxes)
    original_image = Image.open(args.img)
    boxes * = original_image.size* 2
    visualizer.save_image(original_image,boxes,classes,'test1_result.jpg')
    print("The detected image result at {}".format(args.result))
print('Done.')
```

在命令行中输入命令运行测试程序,待测图片如图 6.34 所示,测试结果如图 6.35 所示。

```
python detect.py
```

图 6.34 待测图片

图 6.35 测试结果

小 结

本单元主要阐述了目标检测任务主要的算法发展历程和目标检测的基础知识,分别使用 Yolov3 算法和 SSD 算法完成目标检测的实践任务。

练 习

练习 1　使用 Yolov3-tiny 算法完成 VOC2007 数据集的目标检测任务。
练习 2　使用 SSD 算法完成 VOC2012 数据集的目标检测任务。

单元7 图像分割

图像分割是基于深度学习的图像识别任务之一,是机器视觉技术中关于场景理解的重要一环,是图像处理到图像分析的关键步骤。近年的图像分割技术在各行各业都有着丰富的应用场景。在医学行业,图像分割技术常用于测量医学图像中组织体积和三维重建等;在遥感测控行业,图像分割技术广泛应用于分割合成孔径雷达图像中的目标、提取遥感云图中不同云系与背景、定位卫星图像中的道路和森林等。在新兴的自动驾驶行业,车载相机收集的原始图像都会在汽车核心计算单元上执行图像分割和建模任务,以避让行人和车辆等障碍。随着近些年深度学习的火热,使得图像分割有了巨大的发展。

本单元系统地梳理了图像分割的基础知识,包括传统图像分割方法以及深度学习图像分割方法,希望学生通过本单元的学习能够掌握图像分割的基础知识,并能够熟练使用深度学习框架解决常见的图像分割问题。本单元的知识导图如图7.1所示。

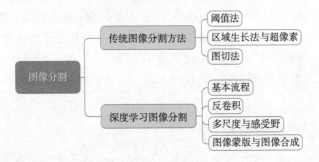

图 7.1 知识导图

课程安排

课程任务	课程目标	安排课时
传统图像分割方法	了解包括阈值法、区域生长法以及图切法在内的传统图像分割方法,了解各类方法的优势与限制性	2
深度学习图像分割方法	掌握基于深度学习的图像分割方法的基本流程,重点掌握全卷积网络的基本结构以及核心技术,并能够熟练运用主流 Mask R-CNN 网络解决图像分割问题	3

7.1 传统图像分割方法

图像分割是图像分析的第一步,是计算机视觉的基础,是图像理解的重要组成部分,也是图像处理中的困难之一。什么是图像分割,顾名思义就是将图像一个个分割成具有特定性质的区域的过程称为图像分割。更完整的定义是指根据灰度、彩色、空间纹理、几何形状等特征把图像划分为若干个互不相交的区域,使得这些特征在同一区域内表现出一致性或相似性,而在不同区域内表现出明显的不同,简言之就是在一幅图像中把目标从背景中分离出来,对于灰度图像来说,区域内部的像素一般具有灰度相似性,而在区域的边界上一般具有灰度不连续性。

图像分割作为图像技术领域的一个经典难题,自20世纪70年代以来吸引了众多研究人员的研究热情并为之付出了巨大努力,提出了很多经典图像分割算法,主要包括阈值法、区域生长与超像素法以及图切法。虽然目前深度学习算法在图像分割中应用广泛,但是依然需要对传统的分割算法有深入理解。

7.1.1 阈值法

阈值分割是常见的直接对图像进行分割的算法,根据图像像素的灰度值的不同而定。对应单一目标图像,只需选取一个阈值,即可将图像分为目标和背景两大类,这个称为单阈值分割;如果目标图像复杂,选取多个阈值,才能将图像中的目标区域和背景分割成多个,这个称为多阈值分割,此时还需要区分检测结果中的图像目标,对各个图像目标区域唯一的标识进行区分。阈值分割的显著优点:成本低廉,实现简单。当目标和背景区域的像素灰度值或其他特征存在明显差异的情况下,该算法能非常有效地实现对图像的分割。阈值分割方法的关键是如何取得一个合适的阈值,常用方法有灰度直方图确定阈值、双峰法选择阈值、迭代选取阈值、大津法选择阈值以及灰度拉伸选择阈值等。

7.1.2 区域生长法与超像素

阈值法仅仅依靠阈值,没有考虑到前景和背景的空间关系。如果考虑到图像的空间关系,则衍生出了区域生长法。区域生长法的基本思想是将有相似性质的像素点合并到一起。对每一个区域要先指定一个种子点作为生长的起点,然后将种子点周围领域的像素点和种子点进行对比,将具有相似性质的点合并起来继续向外生长,直到没有满足条件的像素被包括进来为止。这样一个区域的生长就完成了。

因此区域生长算法一般分为三个步骤实现:

1. 确认种子点

种子点的确认根据具体算法不同,存在差异,但本质上遵循有序随机的规律。假设要确认提取4个种子节点,图像的长和宽为w,h。那么可以先找出四个均匀分布的点$[w/4,h/4]$,$[3w/4,h/4]$,$[w/4,3h/4]$,$[3w/4,3h/4]$,再通过设好的随机规则随机移动四个点。有序性是为了保证分割的平滑,随机性是为了优化分割效果。当然也可以根据需求自行决定种子点。

2. 确认生长准则

一般来说，生长点的选取规则有两种，邻域法和区域法。邻域法如四邻域法、八邻域法等，关注的是像素级的相似度，即尽量纳入与种子像素相似的像素，这种方法对非固定图像的识别更加有效，能很好地找出像素级的相似区域，但对模糊边界的识别较差。区域法则是在纳入新像素时并不只考虑种子像素，而是将已确认区域与其进行整体比较。如图 7.2 所示，如果单考虑种子节点与其邻域，20 与 10 的差距不大，但是如果考虑已有区域的整体灰度值，与 10 的差距就会变得很大。

255	255	255	
255	255	(20)	10
255	255	255	

图 7.2 领域法

总体来说，两种方法各有优缺点，这两种方法的选取一般是根据需要切割的目标决定的。邻域法对差异及其敏感，很容易避开差异较大的位置，但是对模糊边界或梯度变化图像的识别就很差，而区域法能很好地保证切割的整体性和一致性，但是对边界缺乏敏感，很容易陷入欠拟合的问题中。

3. 确认生长停止准则

根据生长准则，一般到达指定大小或所有的点都无法将新的点纳入其中时，停止生长。

区域生长法的关键在于生长种子点和生长准则的确定，其中最经典的方法是分水岭算法。分水岭算法是一种数学形态学方法，如果将二维的图像看作具有高度的地面，图像中每一个像素的灰度值表示海拔高度，那么每一个局部极小值及周围所影响的区域就是"集水盆"，各个集水盆之间相邻，边界就成为了"分水岭"，也就是最终的分割轮廓。

7.1.3 图切法

基于图论的图像分割技术（GraphCut）一直以来都是图像分割领域的研究热点。其基本思想是将图像映射为带权无向图，把像素视作节点，节点之间的边的权重对应于两个像素间的不相似性度量，割的容量对应能量函数。运用最大流/最小流算法对图进行切割，得到的最小割对应于待提取的目标边界。下面介绍基于图论的分割中的 GrabCut 算法。

GrabCut 是常用的一种图像分割算法，该算法是对 GraphCut 的改进和升级，相对于 GraphCut，GrabCut 算法使用迭代估计来训练模型，同时支持不完整的标记，该算法有效地综合了图片中的边界特征和纹理特征，只需要少量的人工交互操作就可以对目标实现较好的分割效果。

GrabCut 使用 k 个高斯分量的全协方差混合高斯模型（GMM）对背景和目标建立模型。混合高斯模型表征图像中每个像素点的特征，图像中的每个像素对应一个高斯分量。在 GraphCut 的基础上，GrabCut 重新定义了 Gibbs 能量 E。能量 E 的最小化代表了一个良好的分割。

GrabCut 算法使用迭代最小化的方式进行算法的更新优化。迭代的方法如下：

① 将图像中目标区域的每一个像素分配到混合高斯模型中，计算其中最大的高斯分量。

②针对得到的高斯分量与像素之间的关系,得到每个高斯模型的均值和协方差参数。
③通过最大流或者最小割算法实现分割能量的最小化。
④重复上述过程,直到算法收敛。

GrabCut算法通过以上迭代得到最小割,从而实现图像的分割。

7.2 深度学习图像分割

在深度学习技术引入计算机视觉领域后获得了巨大成功,尤其是在基于 Convolutional Neural Networks(CNNs)的图像分类、目标检测等领域上效果极佳。这也激励了研究人员探索这类网络能否应用于像素级别的标记问题,比如图像语义分割。下面学习基于深度学习的图像分割方案。

7.2.1 基本流程

全卷积网络(Fully Connected Network,FCN)是研究者于2014年提出的,是最常用的语义分割网络之一,其结构如图7.3所示。

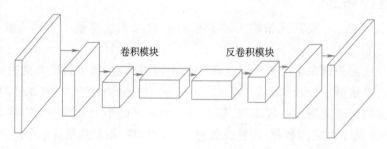

图 7.3 全卷积网络结构

整体的网络结构分为两部分:全卷积部分和反卷积部分。其中全卷积部分借用了一些经典的 CNN 网络(如 AlexNet、VGG、GoogLeNet 等),并把最后的全连接层换成 1×1 卷积,用于提取特征,形成热点图;反卷积部分则是在小尺寸的热点图上采样得到原尺寸的语义分割图像。

作为基于深度学习的图像语义分割的开山之作,目前所有成功的语义分割网络都是基于 FCN 进行改进的。FCN 网络利用 CNNs 在图像上的强大学习能力提出了一个全卷积化概念,将现有的一些常用分类深度网络模型(如 VGG16、GoogLeNet 等网络)的全连接层全部用卷积层替代,这样做的好处是因为没有了全连接层,最终输出的结果是一张图片而不是一维向量,实现了端对端的语义分割;其次,通过去除全连接层能实现任意大小图片的输入,从而保证输入与输出图片大小相等。由于卷积层后接有池化层,池化层又称下采样层,会对图片的分辨率大小产生影响。为了保证输入图片与输出图片的大小相等,FCN 网络使用反卷积的方式进行上采样以维持图片的分辨率。

7.2.2 反卷积

在深入理解基于深度学习的图像分割任务之前,必须掌握反卷积的概念。在应用于计算机视觉的深度学习领域，由于输入图像通过卷积神经网络提取特征后，输出的尺寸往往会变小，而有时我们需要将图像恢复到原来的尺寸以便进行进一步的计算，这个采用扩大图像尺寸，实现图像由小分辨率到大分辨率的映射的操作称为上采样。

上采样有三种常见的方法：双线性插值、反卷积以及反池化，这里只讨论反卷积。这里指的反卷积又称转置卷积，它并不是正向卷积的完全逆过程，用一句话来解释：反卷积是一种特殊的正向卷积，先按照一定的比例通过补0扩大输入图像的尺寸，接着旋转卷积核，再进行正向卷积。

输入图像经过卷积得到的特征图，分辨率明显下降。经过反卷积的上采样操作提升分辨率，同时还保证了特征所在区域的权重，最后将图片的分辨率提升到原图相同尺寸后，权重高的区域则为目标所在区域，如图7.4所示。FCN模型处理过程也是这样，通过卷积和反卷积基本能定位到目标区域。

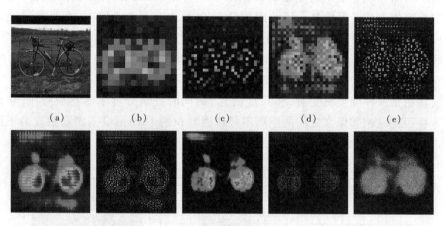

图 7.4　不同方式的上采样

7.2.3　多尺度与感受野

1. 金字塔池化模块

对于分割任务而言，上下文信息的利用情况对于分割的效果是有明显影响的。通常来讲，判断一个东西的类别时，除了直接观察其外观，有时候还会辅助其出现的环境。比如汽车通常出现在道路上、船通常在水面、飞机通常在天上等。忽略了这些直接做判断，有时候就会造成歧义。例如，在水面上的船由于其外观，就被FCN算法判断成了汽车。

对于复杂的场景理解，PSPNet采用空间金字塔池化来融合不同子区域之间的上下文信息。金字塔池化层有如下三个优点：第一，可以解决输入图片大小不一造成的缺陷；第二，可以把一个特征图从不同的角度进行特征提取、聚合；第三，一定程度上提升了模型精度。

具体操作时金字塔池化层在前一卷积层的特征图的每一个图片上进行了3次卷积操作。最右边是原图像，中间是特征图，最右边的就是把图像分成大小是16的特征图。那么每一个feature map 就会变成 16+4+1=21 个 feature maps。即将从这 21 个块中，每个块提取出一个特征，这样刚好就是要提取的21维特征向量。这就解决了特征图大小不一的状况。

采用金字塔池化模型进行特征提取时具体操作如下:模型输入是整张待检测的图片,然后进入 CNN 中,进行一次特征提取,得到特征图,在特征图中找到各个候选框的区域,再对各个候选框采用金字塔空间池化,提取出固定长度的特征向量。

使用金字塔的池化方案可实现不同尺度的感受野,它能够起到将局部区域上下文信息与全局上下文信息结合的效果。对于图像分割任务,全局上下文信息通常是与整体轮廓相关的信息,而局部上下文信息则是图像的细节纹理,要想对多尺度的目标很好地完成分割,这两部分信息都是必需的。

2. 空洞卷积

在图像分割领域,图像输入到 FCN 中,FCN 先像传统的 CNN 那样对图像做卷积再池化,降低图像尺寸的同时增大感受野,但是由于图像分割预测是像素级的输出,所以要将池化后较小的图像尺寸变换到原始的图像尺寸进行预测,之前的池化操作使得每个像素预测都能看到较大感受野信息。因此图像分割 FCN 中有两个关键:一个是通过池化来减小图像尺寸增大感受野;另一个是池化扩大图像尺寸。在先减小再增大尺寸的过程中,肯定有一些信息损失掉了,那么能不能设计一种新的操作,不通过池化也能有较大的感受野,看到更多的信息呢?答案就是空洞卷积。

空洞卷积是一种调整感受野(多尺度信息)的同时控制分辨率的卷积操作,它的原理就是原始卷积区域相邻像素之间的距离不是普通卷积的 1,而是根据膨胀系数的不同而不同,具体操作如图 7.5 所示。

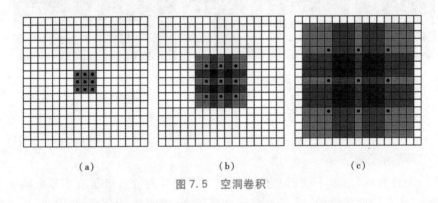

图 7.5 空洞卷积

图 7.5(a)对应 3×3 的 1-空洞卷积,和普通的卷积操作一样,图 7.5(b)对应 3×3 的 2-空洞卷积,实际的卷积核尺寸还是 3×3,但是空洞为 1,也就是对于一个 7×7 的图像块,只有 9 个点和 3×3 的卷积核发生卷积操作,其余的点略过。也可以理解为卷积核的尺寸为 7×7,但是只有图中的 9 个点的权重不为 0,其余都为 0。可以看到虽然卷积核尺寸只有 3×3,但是这个卷积的感受野已经增大到了 7×7(如果考虑到这个 2-空洞卷积的前一层是一个 1-空洞的话,那么每个点就是 1-空洞卷积输出,此时感受野为 3×3,所以 1-空洞卷积和 2-空洞卷积合起来就能达到 7×7 的卷积),图 7.5(c)是 4-空洞卷积操作,同理跟在两个 1-空洞卷积和 2-空洞卷积的后面,能达到 15×15 的感受野。对比传统的卷积操作,3 层 3×3 的卷积加起来,补长为 1 的话,只能达到 7 的感受野,也就是和层数成线性关系,而空洞卷积的感受野呈指数级增长。

很多的研究方法都使用到了空洞卷积,膨胀系数越大,则该卷积核的感受野越大,越有利于综合上下文信息,这对于图像分割、目标检测等任务都非常有效。

7.2.4 图像蒙版与图像合成

1. 图像蒙版

前面所说的图像分割属于硬分割,即每一个像素都以绝对的概率属于某一类,最终概率最大的那一类就是所要的类别。但是这样的分割会带来一些问题,即边缘不够细腻,当后期要进行融合时,边缘过渡不自然。此时,就需要用到图像蒙版(Image Matting)技术。

图像蒙版是通过 alpha 通道控制透明度将图像分为前景图和背景图的技术,以用于后续前景图和新背景图的图像合成,在在线会议、虚拟舞台和视频会议等场景下具有广泛的应用需求。

基于深度学习的图像蒙版方法主要分为三大类,第一类为 Trimap-based 方法,精度高,需要同时输入图像和人工精确标注的 Trimap,比如 DIM;第二类为 Trimap-free 方法,只需要单张图像输入,精度较低,比如 MODNET;第三类为背景蒙版方法,去除标注费时的精确标注的 Trimap 依赖,改为容易获取的"轻微随机"的背景,取得不错的效果。

2. 图像合成

在完成图像分割之后,接下来的一步就是进行背景的替换工作,图像融合是其中的关键技术。从图像中确定前景和背景的技术称为图像蒙版(Image Matting),而将抠出的部分无缝地贴入目标图像的过程则称为图像合成(Image Compositing)。

(1) 直接剪切粘贴

在所有相关技术中,最直观简单的就是直接剪切粘贴,经常在摄影师后期制作中所采用。直接剪切粘贴技术如果应用得当,可以产生相当有艺术感的图像,如图 7.6 所示,这可以认为是在原始的人物和动物身上粘贴了纹理图像形成的。但很多时候,这个简单的技术会产生让人沮丧的结果,生成的照片不够自然。

图 7.6 剪切粘贴法

(2) Alpha 融合

Alpha 融合可以看作一个升级版的直接剪切粘贴,可用如下公式表示:

$$输出 = 前景 \times 蒙版 + 背景 \times (1 - 蒙版)$$

如图 7.7 所示。

图 7.7　Alpha 融合

Alpha Mask 可以利用前面讲述的图像蒙版法获得,合成的图像可能比较不自然。如果对 Mask 图稍微进行羽化,效果就会好很多,对 Mask 图进行羽化的方法很多,最直接的方法就是对其添加一定尺度的高斯滤波。

Alpha 融合的效果取决于如何正确地设置 Alpha Mask,如果要把一张图中的一部分抠出,并融合到另外一张新的图片中,主要牵涉两个步骤:

①准确地抠图获取 Alpha Mask。

②对 Alpha Mask 进行合适的羽化,使得融合更自然。

(3)多频段融合

要想让 Alpha 融合结果显得更自然,关键一点是选择合适的融合窗口大小。以融合两幅图像为例,如图 7.8 和 7.9 所示。

图 7.8　待融合的两幅图

如果选择图像中轴线作为融合后两个图像的分界线,那么融合过程可如图 7.9 表示。

图 7.9　多频段融合

单元任务 17　使用 U-Net 模型实现城市街景图像的分割

U-Net 模型是工程项目中使用最多的图像分割模型,现已成为大多数医疗影像语义分割任务基础的模型结构。因为其网络结构与大写英文字母 U 相像而得名。本任务使用的数据集为城市街景数据集 Cityscape.tar.gz,可从数据资源平台获取。城市街景数据集对于无人驾驶中的场景理解具有重要意义,本项目所用数据集来自 30 多个不同城市街道的高分辨率图像数据,包含 2 975 张训练集和 500 张验证集图片,标签共 34 个类别。目录结构如下,其中 images 文件目录下为图片数据,gtFine 文件目录下为标签数据。本任务使用 tf2.3 开发环境。

```
#下载数据集
wget http://172.16.33.72/dataset/Cityscape.tar.gz
tar -zxvf Cityscape.tar.gz
#数据集目录结构
tree -L 2 dataset
#输出
 dataset
 └gtFine
    ├test
    ├train
    ├val
  └images
    ├test
    ├train
    └val
```

Cityscape 城市街景数据集的原始数据为 1 024×2 048 像素的图片,如图 7.10 所示。使用的标签为未涂色的 mask 图片,在 gtFine 中以_gtFine_labelIds.png 名称结尾的图片,如图 7.11 所示。将 mask 标签涂色后的图片显示如图 7.12 所示。

图 7.10　原始图片

图 7.11 未涂色的 mask 标签

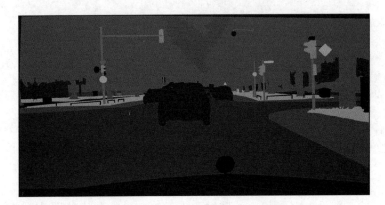

图 7.12 涂色的 mask 标签

步骤 1：数据读取和预处理。

创建 dataset.py 文件，分别对训练集和验证集执行数据读取、数据预处理和载入数据管道操作。

```python
import tensorflow as tf
import glob
import numpy as np
import matplotlib.pyplot as plt

TRAIN_IMAGE_PATH = "dataset/images/train/*/*_leftImg8bit.png"
TRAIN_MASK_PATH = "dataset/gtFine/train/*/*_gtFine_labelIds.png"
VAL_IMAGE_PATH = "dataset/images/val/*/*_leftImg8bit.png"
VAL_MASK_PATH = "dataset/gtFine/val/*/*_gtFine_labelIds.png"
AUTOTUNE = tf.data.experimental.AUTOTUNE

def read_image(path):
    #读取图片
    image = tf.io.read_file(path)
    image = tf.image.decode_png(image, channels=3)
    return image
```

```python
def read_mask(path):
    #读取标签
    mask = tf.io.read_file(path)
    mask = tf.image.decode_png(mask,channels=1)
    return mask

def preprocessing_dataset(image,mask):
    #数据预处理
    #将iamge和mask按通道合并
    concat_image_mask = tf.concat([image,mask],axis=-1)
    #改变尺寸
    concat_image_mask = tf.image.resize(concat_image_mask,(280,280),method=tf.image.ResizeMethod.NEAREST_NEIGHBOR)
    #随机裁剪
    concat_image_mask = tf.image.random_crop(concat_image_mask,[256,256,4])
    #拆分image和mask并返回
    return concat_image_mask[:,:,:3],concat_image_mask[:,:,3:]

def normal(image,mask):
    #归一化
    image = tf.cast(image,dtype=tf.float32) / 127.5-1
    mask = tf.cast(mask,tf.int32)
    return image,mask

def load_train(image_path,mask_path):
    #处理train数据
    image = read_image(image_path)
    mask = read_mask(mask_path)
    image,mask = preprocessing_dataset(image,mask)
    image,mask = normal(image,mask)
    return image,mask

def load_val(image_path,mask_path):
    #处理val数据
    image = read_image(image_path)
    mask = read_mask(mask_path)
    image = tf.image.resize(image,(256,256))
    mask = tf.image.resize(mask,(256,256))
    image,mask = normal(image,mask)
    return image,mask

def load_aug(image_path,mask_path):
    #载入数据
    image = read_image(image_path)
    mask = read_mask(mask_path)
    image,mask = preprocessing_dataset(image,mask)
    image_filp = tf.image.flip_left_right(image)
    mask_flip = tf.image.flip_left_right(mask)
```

```python
        image_aug = tf.concat([image, image_flip], axis = 0)
        mask_aug = tf.concat([mask, mask_flip], axis = 0)
        image_aug, mask_aug = normal(image_aug, mask_aug)
        return image_aug, mask_aug

    def load_ds_pipeline(batch_size):
        #载入数据为数据管道
        #获取数据集图片路径和标签
        train_image_path = glob.glob(TRAIN_IMAGE_PATH)
        train_mask_path = glob.glob(TRAIN_MASK_PATH)
        val_image_path = glob.glob(VAL_IMAGE_PATH)
        val_mask_path = glob.glob(VAL_MASK_PATH)
        #排序
        train_image_path.sort()
        train_mask_path.sort()
        val_image_path.sort()
        val_mask_path.sort()
        #train 数量和 val 数量
        num_train = len(train_image_path)
        num_val = len(val_image_path)
        #按同一顺序打乱数据和标签
        random_train_index = np.random.permutation(num_train)
        train_image_path = np.array(train_image_path)[random_train_index]
        train_mask_path = np.array(train_mask_path)[random_train_index]
        random_val_index = np.random.permutation(num_val)
        val_image_path = np.array(val_image_path)[random_val_index]
        val_mask_path = np.array(val_mask_path)[random_val_index]
        #训练集数据管道
        train_ds = tf.data.Dataset.from_tensor_slices((train_image_path, train_mask_path))
        train_ds = train_ds.map(load_train, num_parallel_calls = AUTOTUNE)
        train_ds = train_ds.shuffle(buffer_size = 500).batch(batch_size)
        train_ds = train_ds.prefetch(buffer_size = AUTOTUNE)
        #验证集数据管道
        val_ds = tf.data.Dataset.from_tensor_slices((val_image_path, val_mask_path))
        val_ds = val_ds.map(load_val, num_parallel_calls = AUTOTUNE)
        val_ds = val_ds.batch(batch_size)
        val_ds = val_ds.prefetch(buffer_size = AUTOTUNE)

        return train_ds, val_ds

    def load_step(batch_size):
        #加载训练集和验证集步数
        train_image_path = glob.glob(TRAIN_IMAGE_PATH)
        val_image_path = glob.glob(VAL_IMAGE_PATH)
        num_train = len(train_image_path)
        num_val = len(val_image_path)
```

```
            steps_per_epoch = num_train //batch_size
            validation_steps = num_val //batch_size
            return steps_per_epoch,validation_steps
```

步骤2:构建 U-Net 模型。

U-Net 模型结构如图 7.13 所示。

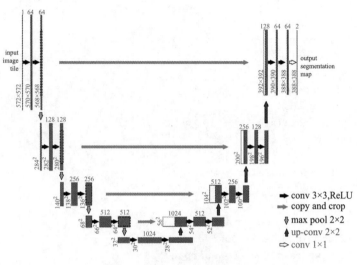

图 7.13　U-Net 模型结构

根据图 7.13 可以自定义 tf.keras.Model 类构建 U-net 模型,创建模型文件 Unet.py,写入模型代码。

```
import tensorflow as tf

class Downsample(tf.keras.layers.Layer):
    #自定义下采样
    def __init__(self,units):
        super(Downsample,self).__init__()
        self.conv1 = tf.keras.layers.Conv2D(units,kernel_size = 3,padding = 'same')
        self.conv2 = tf.keras.layers.Conv2D(units,kernel_size = 3,padding = 'same')
        self.pool = tf.keras.layers.MaxPooling2D(padding = 'same')

    def call(self,x,use_pool = True):
        if use_pool:
            x = self.pool(x)
        x = self.conv1(x)
        x = tf.nn.relu(x)
        x = self.conv2(x)
        y = tf.nn.relu(x)
        return y

class Upsample(tf.keras.layers.Layer):
    #自定义上采样
```

```python
    def __init__(self,units):
        super(Upsample,self).__init__()
        self.conv1 = tf.keras.layers.Conv2D(units,kernel_size =3,padding = 'same')
        self.conv2 = tf.keras.layers.Conv2D(units,kernel_size =3,padding = 'same')
        self.deconv = tf.keras.layers.Conv2DTranspose(units//2,kernel_size =2,strides =2,padding = 'same')

    def call(self,x):
        x = self.conv1(x)
        x = tf.nn.relu(x)
        x = self.conv2(x)
        x = tf.nn.relu(x)
      x = self.deconv(x)
        y = tf.nn.relu(x)
        return y

class Unet(tf.keras.Model):
    #自定义 Unet
    def __init__(self):
        super(Unet,self).__init__()
        self.down_sample_1 = Downsample(64)
        self.down_sample_2 = Downsample(128)
        self.down_sample_3 = Downsample(256)
        self.down_sample_4 = Downsample(512)
        self.down_sample_5 = Downsample(1024)
        self.up_sample_0 = tf.keras.layers.Conv2DTranspose(512,kernel_size =2,strides =2,padding = 'same')
        self.up_sample_1 = Upsample(512)
        self.up_sample_2 = Upsample(256)
        self.up_sample_3 = Upsample(128)
        self.conv_last = Downsample(64)
        self.outputs = tf.keras.layers.Conv2D(34,kernel_size =1,padding = 'same')

    def call(self,x):
        x1 = self.down_sample_1(x,use_pool = False)
        x2 = self.down_sample_2(x1)
        x3 = self.down_sample_3(x2)
        x4 = self.down_sample_4(x3)
        x5 = self.down_sample_5(x4)

        x5 = self.up_sample_0(x5)
        x5 = tf.concat([x4,x5],axis = -1)
        x5 = self.up_sample_1(x5)
        x5 = tf.concat([x3,x5],axis = -1)
        x5 = self.up_sample_2(x5)
        x5 = tf.concat([x2,x5],axis = -1)
        x5 = self.up_sample_3(x5)
        x5 = tf.concat([x1,x5],axis = -1)
```

```python
        x5 = self.conv_last(x5, use_pool = False)
        outputs = self.outputs(x5)
    return outputs

#子类化 tf.keras.metrics.MeanIoU
class MeanIoU(tf.keras.metrics.MeanIoU):
    def __call__(self, y_true, y_pred, sample_weight = None):
        y_pred = tf.argmax(y_pred, axis = -1)
        return super().__call__(y_true, y_pred, sample_weight = sample_weight)
```

步骤3：训练模型。

创建训练程序train.py，使用自定义循环的方式训练模型，先定义一个批量的处理方式，再通过迭代循环，最后保存模型的权重参数。

```python
import os
import tensorflow as tf
from dataset import load_ds_pipeline, load_step
from Unet import Unet, MeanIoU
from tqdm import tqdm

os.environ['CUDA_VISIBLE_DEVICES'] = '0'
os.environ['TF_FORCE_GPU_ALLOW_GROWTH'] = 'true'
os.environ['TF_CPP_MIN_LOG_LEVEL'] = '3'
BATCH_SIZE = 2
EPOCH = 30
MODEL_PATH = "data/epoch30_model_weights.h5"

#载入数据
print("加载数据>>>")
train_ds, val_ds = load_ds_pipeline(BATCH_SIZE)
steps_per_epoch, _ = load_step(BATCH_SIZE)

#定义更新的评估指标
train_loss = tf.keras.metrics.Mean(name = 'train_loss')
train_accuracy = tf.keras.metrics.SparseCategoricalAccuracy(name = 'train_accuracy')
train_iou = MeanIoU(34, name = 'train_iou')

val_loss = tf.keras.metrics.Mean(name = 'val_loss')
val_accuracy = tf.keras.metrics.SparseCategoricalAccuracy(name = 'val_accuracy')
val_iou = MeanIoU(34, name = 'val_iou')

#构建模型
print("加载模型>>>")
model = Unet()
#定义优化器
optimizer = tf.keras.optimizers.Adam(0.0001)
#定义损失函数
loss_fn = tf.keras.losses.SparseCategoricalCrossentropy(from_logits = True)
```

```python
@tf.function
def train_step(image,mask):
    #train
    with tf.GradientTape() as tape:
        prediction = model(image)
        loss = loss_fn(mask,prediction)
    gradient = tape.gradient(loss,model.trainable_variables)
    optimizer.apply_gradients(zip(gradient,model.trainable_variables))
    train_loss(loss)
    train_accuracy(mask,prediction)
    train_iou(mask,prediction)

@tf.function
def val_step(image,mask):
    #val
    prediction = model(image)
    loss = loss_fn(mask,prediction)

    val_loss(loss)
    val_accuracy(mask,prediction)
    val_iou(mask,prediction)

print("开始训练>>>")
for epoch in range(EPOCH):
    #训练过程
    train_loss.reset_states()
    train_accuracy.reset_states()
    train_iou.reset_states()
    val_loss.reset_states()
    val_accuracy.reset_states()
    val_iou.reset_states()

    for image,mask in tqdm(train_ds,total = steps_per_epoch):
        train_step(image,mask)

    for image,mask in val_ds:
        val_step(image,mask)

    template = "Epoch {}: train_loss: {:.3f}; train_accuracy: {:.3%}; train_IOU: {:.3f}; \
        val_loss: {:.3f}; val_accuracy: {:.3%}; val_IOU: {:.3f}"
    print(template.format(epoch + 1, train_loss.result(), train_accuracy.result(),train_iou.result(),
        val_loss.result(),val_accuracy.result(),val_iou.result()))

print("训练完成!")
#保存模型参数
model.save_weights(MODEL_PATH)
```

运行训练程序,如图 7.14 和图 7.15 所示。

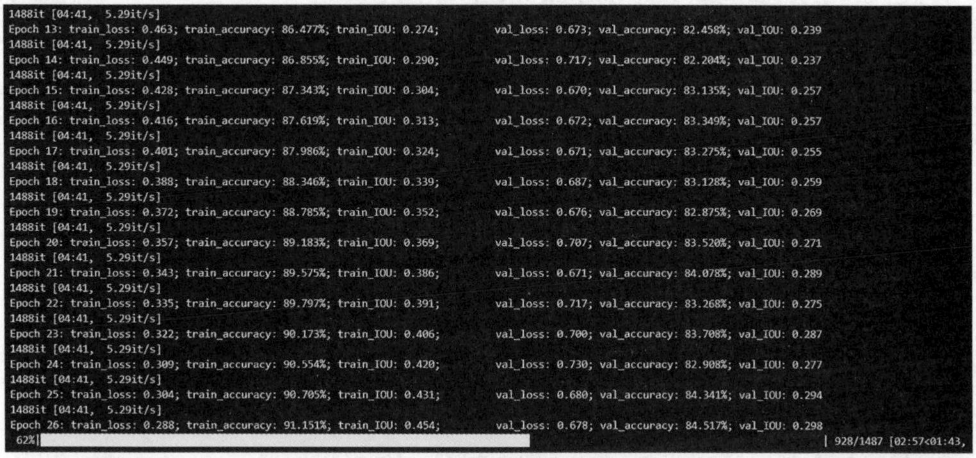

图 7.14　训练模型(一)

图 7.15　训练模型(二)

步骤 4：评估模型。

在模型评估 eval.py 中,直接使用验证集数据进行评估,重载模型后分别计算验证集的准确率和 MeanIoU 值。

```
import os
import tensorflow as tf
from dataset import load_ds_pipeline,load_step
from Unet import Unet,MeanIoU
from tqdm import tqdm

os.environ['CUDA_VISIBLE_DEVICES'] = '0'
os.environ['TF_FORCE_GPU_ALLOW_GROWTH'] = 'true'
os.environ['TF_CPP_MIN_LOG_LEVEL'] = '3'
BATCH_SIZE = 1
MODEL_PATH = "data/epoch30_model_weights.h5"
```

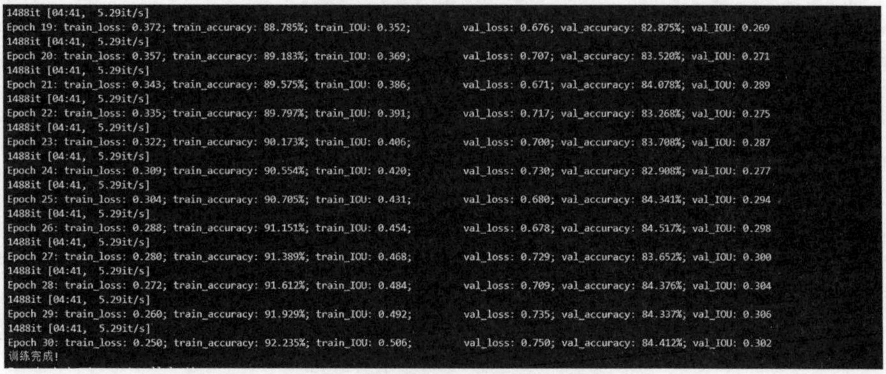

```python
print("加载数据>>>")
_,val_ds = load_ds_pipeline(BATCH_SIZE)
_,validation_steps = load_step(BATCH_SIZE)

print("重载模型>>>")
inputs = tf.keras.Input(shape = (256,256,3))
model = Unet()
model(inputs)
optimizer = tf.keras.optimizers.Adam(0.0001)
loss_fn = tf.keras.losses.SparseCategoricalCrossentropy(from_logits = True)
model.compile(optimizer = optimizer,loss = loss_fn,metrics = [
    tf.keras.metrics.SparseCategoricalAccuracy(),MeanIoU(34)
])
model.load_weights(MODEL_PATH)

val_loss = tf.keras.metrics.Mean(name = 'val_loss')
val_accuracy = tf.keras.metrics.SparseCategoricalAccuracy(name = 'val_accuracy')
val_iou = MeanIoU(34,name = 'val_iou')

@tf.function
def val_step(image,mask):
    #val
    prediction = model(image)
    loss = loss_fn(mask,prediction)

    val_loss(loss)
    val_accuracy(mask,prediction)
    val_iou(mask,prediction)

print("开始评估>>>")
for image,mask in tqdm(val_ds,total = validation_steps):
    val_step(image,mask)

print("评估完成! \nAccuracy: {:.3%}; MeanIoU: {:.3f}".format(val_accuracy.result(),
val_iou.result()))
```

运行评估程序,如图7.16所示。

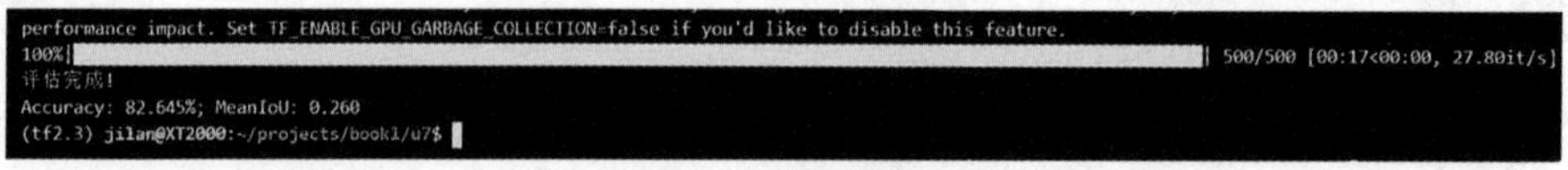

图7.16 评估模型

可知模型在验证集上的34类语义分割的准确率达到82.645%,平均交并比得分0.260。

步骤5:测试模型。

创建segment.py文件,重载保存的模型对任意一张测试集图片进行语义分割,返回U-Net的分割掩码图并涂色显示。

```python
import os
import tensorflow as tf
import numpy as np
import matplotlib.pyplot as plt
from Unet import Unet,MeanIoU

os.environ['CUDA_VISIBLE_DEVICES'] = '0'
os.environ['TF_FORCE_GPU_ALLOW_GROWTH'] = 'true'
os.environ['TF_CPP_MIN_LOG_LEVEL'] = '3'
IMAGE_PATH = ""
MODEL_PATH = "data/epoch30_model_weights.h5"
RESULT_PATH = "data/result.png"

print("图片预处理>>>")
image = tf.io.read_file(IMAGE_PATH)
image = tf.image.decode_png(image,channels=3)
image = tf.image.resize(image,(256,256))
image = tf.cast(image,dtype=tf.float32) / 127.5 - 1
image = tf.expand_dims(image,axis=0)

print("重载模型>>>")
inputs = tf.keras.Input(shape=(256,256,3))
model = Unet()
model(inputs)
optimizer = tf.keras.optimizers.Adam(0.0001)
loss_fn = tf.keras.losses.SparseCategoricalCrossentropy(from_logits=True)
model.compile(optimizer=optimizer,loss=loss_fn,metrics=[
    tf.keras.metrics.SparseCategoricalAccuracy(),MeanIoU(34)
])
model.load_weights(MODEL_PATH)

print("模型预测>>>")
mask = model(image)
mask = tf.argmax(mask,axis=-1)
print("完成!")

plt.figure()
plt.subplot(1,2,1)
plt.axis('off')
plt.title('image')
plt.imshow(np.squeeze((image.numpy()+1)/2))
plt.subplot(1,2,2)

plt.axis('off')
plt.title('mask')
plt.imshow(np.squeeze(mask.numpy()))
plt.savefig(RESULT_PATH)
```

运行 segment.py 文件,对一张测试集图片进行语义分割掩码图并涂色,显示结果如图 7.17 所示。

图 7.17 测试图和涂色的掩码图

任务总结:本单元中完成了一个图像分割任务,基于预训练模型对简单的图像取得了不错的结果,模型仍然有较大改进空间。通过一个分割模型的改进,可以从图像数据增强、提高分辨率和改进网络的大小等方面着手,读者可以去完成更多的实验并尝试在自己的数据上构建模型。

小　　结

图像分割可作为预处理将最初的图像转化为若干个更加抽象、更便于计算机处理的形式,既保留了图像中的重要特征信息,又有效地减少了图像中的无用数据、提高了后续图像处理的准确率和效率。在通信方面,可事先提取目标的轮廓结构、区域内容等,保证不失有用信息的同时,有针对性地压缩图像,以提高网络传输效率;在交通领域可用来对车辆进行轮廓提取、识别或跟踪,行人检测等。总的来说,凡是与目标的检测、提取和识别等相关的内容,都需要利用到图像分割技术。因此,无论是从图像分割的技术和算法,还是从对图像处理、计算机视觉的影响以及实际应用等各个方面来深入研究和探讨图像分割,都具有十分重要的意义。

练　　习

使用 U-Net 模型实现对核磁共振图像的分割。

阿尔茨海默病又称老年痴呆症,现如今人口老年化的趋势愈加明显,老年人普遍会受到老年痴呆症的困扰。据研究,它是一种由脑组织海马体退化所引起的严重疾病,因此就可以通过脑部核磁共振成像(MRI)和图像分割技术对人脑海马体进行研究来预测患者患病的可能性。这是一个图像分割技术在医疗卫生领域内成功的典型案例。本练习所使用的数据集是脑部核磁共振图像,可以从数据资源平台下载。该数据集包含训练集和测试集两个子集,训练集中包含有 100 位患者的脑部核磁共振图和标注图,测试集中包含有 35 位患者的脑部核磁共振图和标注图。

单元8 图像生成

深度学习在分类任务上取得了革命性的突破,但是需要大量的有标签数据作为支撑。当数据匮乏的时候,神经网络极易出现过拟合问题,这种现象在小规模数据集上尤为明显。传统的图像领域的数据增强技术是建立在一系列已知的仿射变换(如旋转、缩放、位移等),以及一些简单的图像处理手段(如光照色彩变换、对比度变换、添加噪声等)基础上的。这些变化的前提是不改变图像的标签,并且只能局限在图像领域。这种基于几何变换和图像操作的数据增强方法可以在一定程度上缓解神经网络过拟合的问题,提高泛化能力。但是相比于原始数据而言,增加的数据点并没有从根本上解决数据不足的难题;同时,这种数据增强方式需要人为设定转换函数和对应的参数,一般都是凭借经验知识,最优数据增强通常难以实现,所以模型的泛化性能只能得到有限的提升。

图像生成普遍使用生成对抗模型(GAN),生成对抗模型是一种无监督算法,由于其出色的性能引起了人们的广泛关注。这种基于网络合成的方法相比于传统的数据增强技术虽然过程更加复杂,但是生成的样本更加多样,同时还可以应用于图像编辑、图像去噪和图像增强等场景。本单元主要介绍的是基于生成对抗网络的图像生成技术,并将这种方法应用于小规模数据集的图像增广任务。本单元的知识导图如图8.1所示。

图8.1 知识导图

课程安排

课程任务	课程目标	安排课时
掌握生成对抗网络基础知识	学习包括生成对抗网络的算法思想、经典网络结构,了解生成式模型与判别式模型,进一步理解生成器和判别器原理,通过实战掌握生成对抗网络的构建及训练方式	4

8.1 生成对抗网络基础

2014年,生成对抗网络(Generative Adversarial Network,GAN)提出以来,掀起了一股研究热潮。GAN由生成器和判别器组成,生成器负责生成样本,判别器负责判断生成器生成的样本是否为真。生成器要尽可能迷惑判别器,而判别器要尽可能区分生成器生成的样本和真实样本。

完整的生成对抗网络包含两部分,分别为生成器和判别器。生成器就像造假货的犯罪分子,判别器就如同警察。犯罪分子努力让假货看起来逼真,警察则不断提升对于假货的辨识能力。二者互相博弈,随着时间的进行,都会越来越强。在图像生成任务中,生成器不断生成尽可能逼真的假图像。判别器则判断图像是否是真实的图像,二者在比较中不断博弈优化。最终生成器生成的图像使得判别器完全无法判别真假。

通常情况下,无论是生成器还是判别器,都可以用神经网络实现。那么,可以把通俗化的定义用图8.2所示的模型表示。

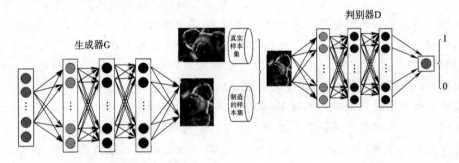

图 8.2 生成器和判别器

上述模型左边是生成器 G,其输入是 z,对于原始的 GAN,z 是由高斯分布随机采样得到的噪声。噪声 z 通过生成器 G 得到了生成的假样本。

生成的假样本与真实样本放到一起,被随机抽取送入到判别器 D,由判别器区分输入的样本是生成的假样本还是真实的样本。整个过程简单明了,生成对抗网络中的"生成对抗"主要体现在生成器和判别器之间的对抗。

8.2 生成对抗网络结构

生成对抗网络包含了两个子网络:生成网络(Generator,简称 G)和判别网络(Discriminator,简称 D),其中生成网络 G 负责学习样本的真实分布,判别网络 D 负责将生成网络采样的样本与真实样本区分开来。

8.2.1 生成网络

生成网络 G 和自编码器的 Decoder 功能类似,从先验分布 $p_z(\cdot)$ 中采样隐藏变量 $z \sim p_z(\cdot)$,通过生成网络 G 参数化的 $p_g(x|z)$ 分布,获得生成样本 $x \sim p_g(x|z)$,如图8.3所示。其中隐藏变量 z 的先验分布 $p_z(\cdot)$ 可以假设为某种已知的分布,比如多元均匀分布 $z \sim \text{Uniform}(-1,1)$。

$p_g(x|z)$ 可以用深度神经网络来参数化,如图 8.4 所示,从均匀分布 $p_z(\cdot)$ 中采样出隐藏变量 z,经过多层转置卷积层网络参数化的 $p_g(x|z)$ 分布中采样出样本 x_f。从输入/输出层面来看,生成器 G 的功能是将隐藏向量 z 通过神经网络转换为样本向量 x_f,下标 f 代表假样本。

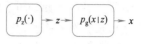

图 8.3 生成模型

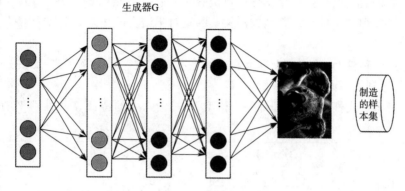

图 8.4 深度生成模型

8.2.2 判别网络

判别网络和普通的二分类网络功能类似,它接受输入样本 x 的数据集,包含了采样自真实数据分布 $p_r(\cdot)$ 的样本 $x_r \sim p_r(\cdot)$,也包含了采样自生成网络的假样本 $x_f \sim p_g(x|z)$,x_r 和 x_f 共同组成了判别网络的训练数据集。判别网络输出为 x 属于真实样本的概率 $P(x\text{为真}|x)$,把所有真实样本 xr 的标签标注为真(1),所有生成网络产生的样本 x_f 标注为假(0),通过最小化判别网络 D 的预测值与标签之间的误差来优化判别网络参数,如图 8.5 所示。

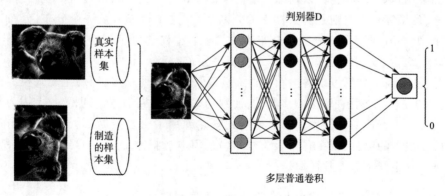

图 8.5 判别模型

8.3 生成式模型与判别式模型

"生成器"和"判别器"的概念其实并非在生成对抗网络中首次提出,在机器学习模型中早已有之。对于机器学习模型,可以根据模型对数据的建模方式将模型分为两大类,生成式模型和判别式模型。如果要训练一个关于猫狗分类的模型,对于判别式模型,只需要学习二者的差异即可。比如说猫的体型会比狗小一点。而生成式模型则不一样,需要学习猫长什么样,狗长什么样。有了二者的长相以后,再根据长相去区分。具体而言:

- 生成式模型:由数据学习联合概率分布 $P(X,Y)$,然后由 $P(Y|X) = P(X,Y)/P(X)$ 求出概率分布 $P(Y|X)$ 作为预测的模型。该方法表示了给定输入 X 与产生输出 Y 的生成关系
- 判别式模型:由数据直接学习决策函数 $Y = f(X)$ 或条件概率分布 $P(Y|X)$ 作为预测模型,即判别模型。判别方法关心的是对于给定的输入 X,应该预测什么样的输出 Y。

对于上述两种模型,从文字上理解起来似乎不太直观。举个例子来阐述一下,对于性别分类问题,分别用不同的模型来做:

①如果用生成式模型:可以训练一个模型,学习输入人的特征 X 和性别 Y 的关系。比如现在有下面一批数据:

	Y(性别)	0	1
X(特征)	0	1/4	3/4
X(特征)	1	3/4	1/4

这个数据可以统计得到,即统计人的特征 $X = 0,1$ 时,其类别为 $Y = 0,1$ 的概率。统计得到上述联合概率分布 $P(X,Y)$ 后,可以学习一个模型,比如让二维高斯分布去拟合上述数据,这样就学习到了 X,Y 的联合分布。在预测时,如果希望给一个输入特征 X,预测其类别,则需要通过贝叶斯公式得到条件概率分布才能进行推断。

$$P(Y|X) = \frac{P(X,Y)}{P(X)} = \frac{P(X,Y)}{P(X|Y)P(Y)}$$

②如果用判别式模型:可以训练一个模型,输入人的特征 X,这些特征包括人的五官、穿衣风格、发型等。输出则是对于性别的判断概率,这个概率服从一个分布,分布的取值只有两个,要么男,要么女,记这个分布为 Y。这个过程学习了一个条件概率分布 $P(Y|X)$,即输入特征 X 的分布已知条件下,Y 的概率分布。

显然,从上面的分析可以看出。判别式模型似乎要方便很多,因为生成式模型要学习一个 X,Y 的联合分布往往需要很多数据,而判别式模型需要的数据则相对少,因为判别式模型更关注输入特征的差异性。不过生成式既然使用了更多数据来生成联合分布,自然也能够提供更多的信息,现在有一个样本 (X,Y),其联合概率 $P(X,Y)$ 经过计算特别小,那么可以认为这个样本是异常样本。

单元任务 18 代码实现 GAN 算法生成人脸图片

本任务完成一个人脸图片生成实战,本任务的主要目标是使用小型人脸数据集训练出可以自动生成人脸的 GAN 模型。本项目使用 tf1.15 开发环境。相关人脸数据集可以从数据资源平台获取。下载好数据并解压,数据的所有文件的格式均为 *.jpg,图像的尺寸大小均为 256×256×3,真彩色影像,且每张图片均含有不同状态的人脸信息。数据集目录如下。

```
face(文件夹)
 ├─A（A 的人脸数据集）
 ├─image1.jpg
 ├─image2.jpg
 ├─image3.jpg
 ├─...
```

步骤 1：数据预处理。

导入模型需要的库，并定义相关参数。

```python
#导入需要的包
from PIL import Image              #Image 用于读取影像
from skimage import io             #io 也可用于读取影像,效果比 Image 读取的更好一些

import tensorflow as tf            #用于构建神经网络模型
import matplotlib.pyplot as plt    #用于绘制生成影像的结果
import numpy as np                 #读取影像
import os                          #文件夹操作
import time                        #计时

#设置相关参数
is_training = True
input_dir = "./face/"              #原始数据的文件夹路径

#设置超参数 hyper parameters
batch_size = 64
image_width = 64
image_height = 64
image_channel = 3
data_shape = [64,64,3]
data_length = 64 * 64 * 3

z_dim = 100
learning_rate = 0.00005
beta1 = 0.5
epoch = 500
```

原始数据无法进行处理,需要将其读到程序中,读取图像的代码如下：

```python
#读取数据的函数
def prepare_data(input_dir, floder):
    '''
    函数功能：通过输入图像的路径,读取训练数据
    参数 input_dir：图像数据所在的根目录,即"./face"
    参数 floder：图像数据所在的子目录,即"./face/A"
    return：返回读取好的训练数据
    '''

    #遍历图像路径,并获取图像数量
    images = os.listdir(input_dir + floder)
    image_len = len(images)
```

```python
#设置一个空 data,用于存放数据
data = np.empty((image_len,image_width,image_height,image_channel),dtype = "float32")

#逐个图像读取
for i in range(image_len):
    #如果导入的是 skimage.io,则读取影像应该写为 img = io.imread(input_dir + images[i])
    img = Image.open(input_dir + floder + "/" + images[i])    #打开图像
    img = img.resize((image_width,image_height))              #将 256×256 变成 64×64
    arr = np.asarray(img,dtype = "float32")                   #将格式改为 np.array
    data[i,:,:,:] = arr                                       #将其放入 data 中

sess = tf.Session()
sess.run(tf.initialize_all_variables())
data = tf.reshape(data,[-1,image_width,image_height,image_channel])
train_data = data* 1.0 / 127.5 - 1.0                          #对 data 进行正则化
train_data = tf.reshape(train_data,[-1,data_length])          #将其拉伸成一维向量
train_set = sess.run(train_data)
sess.close()
return train_set
```

步骤 2:构建 GAN 模型。

按照 GAN 的思路,网络结构的核心是生成器和判别器,同时再加上存储函数即可。

定义生成器函数,代码如下:

```python
#定义生成器
def Generator(z,is_training,reuse):
    '''
    函数功能:输入噪声 z,生成图像 gen_img
    param z:即输入数据,一般为噪声
    param is_training:是否为训练环节
    return:返回生成影像 gen_img
    '''
    
    #图像的 channel 维度变化为 1 - >1024 - >512 - >256 - >128 - >3
    depths = [1024,512,256,128] + [data_shape[2]]
    
    with tf.variable_scope("Generator",reuse = reuse):
        #第一层全连接层
        with tf.variable_scope("g_fc1",reuse = reuse):
            output = tf.layers.dense(z,depths[0]* 4* 4,trainable = is_training)
            output = tf.reshape(output,[batch_size,4,4,depths[0]])
            output = tf.nn.relu(tf.layers.batch_normalization(output,training = is_training))
        
        #第二层反卷积层 1024
        with tf.variable_scope("g_dc1",reuse = reuse):
```

```python
            output = tf.layers.conv2d_transpose(output,depths[1],[5,5],strides = 
(2,2),padding = "SAME",trainable = is_training)
            output = tf.nn.relu(tf.layers.batch_normalization(output,training = 
is_training))

        #第三层反卷积层 512
        with tf.variable_scope("g_dc2",reuse = reuse):
            output = tf.layers.conv2d_transpose(output,depths[2],[5,5],strides = (2,2),
padding = "SAME",trainable = is_training)
            output = tf.nn.relu(tf.layers.batch_normalization(output,training = is_
training))

        #第四层反卷积层 256
        with tf.variable_scope("g_dc3",reuse = reuse):
            output = tf.layers.conv2d_transpose(output,depths[3],[5,5],strides = 
(2,2),padding = "SAME",trainable = is_training)
            output = tf.nn.relu(tf.layers.batch_normalization(output,training = is_
training))

        #第五层反卷积层 128
        with tf.variable_scope("g_dc4",reuse = reuse):
            output = tf.layers.conv2d_transpose(output, depths[4],[5,5],
strides = (2,2),padding = "SAME",trainable = is_training)
            gen_img = tf.nn.tanh(output)

    return gen_img
```

编写判别器函数,代码如下:

```python
#定义判别器
def Discriminator(x,is_training,reuse):
    '''
    函数功能:判别输入的图像是真或假
    param x: 输入数据
    param is_training: 是否为训练环节
    return: 判别结果
    '''

    #channel 维度变化为:3 - >64 - >128 - >256 - >512
    depths = [data_shape[2]] + [64,128,256,512]

    with tf.variable_scope("Discriminator",reuse = reuse):
        #第一层卷积层,注意用的是 leaky_relu 函数
        with tf.variable_scope("d_cv1",reuse = reuse):
            output = tf.layers.conv2d(x,depths[1],[5,5],strides = (2,2),padding = "
SAME",trainable = is_training)
            output = tf.nn.leaky_relu(tf.layers.batch_normalization
(output,training = is_training))
```

```python
            #第二层卷积层,注意用的是leaky_relu函数
            with tf.variable_scope("d_cv2",reuse = reuse):
                output = tf.layers.conv2d(output,depths[2],[5,5],strides = (2,2),padding = "SAME",trainable = is_training)
                output = tf.nn.leaky_relu(tf.layers.batch_normalization(output,training = is_training))

            #第三层卷积层,注意用的是leaky_relu函数
            with tf.variable_scope("d_cv3",reuse = reuse):
                output = tf.layers.conv2d(output,depths[3],[5,5],strides = (2,2),padding = "SAME",trainable = is_training)
                output = tf.nn.leaky_relu(tf.layers.batch_normalization(output,training = is_training))

            #第四层卷积层,注意用的是leaky_relu函数
            with tf.variable_scope("d_cv4",reuse = reuse):
                output = tf.layers.conv2d(output,depths[4],[5,5],strides = (2,2),padding = "SAME",trainable = is_training)
                output = tf.nn.leaky_relu(tf.layers.batch_normalization(output,training = is_training))

            #第五层全链接层
            with tf.variable_scope("d_fc1",reuse = reuse):
                output = tf.layers.flatten(output)
                disc_img = tf.layers.dense(output,1,trainable = is_training)

        return disc_img
```

编写保存结果的函数,代码如下:

```python
def plot_and_save(order,images):
    '''
    函数功能:绘制生成器的结果,并保存
    param order:文件名
    param images:输入图片
    '''

    #将一个batch_size的所有图像进行保存
    batch_size = len(images)
    n = np.int(np.sqrt(batch_size))

    #读取图像大小,并生成掩模canvas
    image_size = np.shape(images)[2]
    n_channel = np.shape(images)[3]
    images = np.reshape(images,[-1,image_size,image_size,n_channel])
    canvas = np.empty((n* image_size,n* image_size,image_channel))

    #为每个掩模赋值
    for i in range(n):
```

```python
            for j in range(n):
                canvas[i*image_size:(i+1)*image_size,j*image_size:(j+1)*image_size,:] = images[n*i+j].reshape(64,64,3)

    #绘制结果,并设置坐标轴
    plt.figure(figsize = (8,8))
    plt.imshow(canvas,cmap = "gray")
    label = "Epoch: {0}".format(order +1)
    plt.xlabel(label)

    #为每个文件命名
    if type(order) is str:
        file_name = order
    else:
        file_name = "face_gen" + str(order)

    #保存绘制的结果
    plt.savefig(file_name)
    print(os.getcwd())
    print("Image saved in file: ",file_name)
    plt.close()
```

步骤3:模型训练与评估。

定义训练过程的函数,代码如下:

```python
#定义训练过程
def training():
    '''
    函数功能:实现DCGAN的训练过程
    '''
    #准备数据。这里输入根目录,以A的影像为例进行图像生成
    data = prepare_data(input_dir,"A")

    #构建网络结构
    x = tf.placeholder(tf.float32,shape = [None,data_length],name = "Input_data")
    x_img = tf.reshape(x,[-1] + data_shape)
    z = tf.placeholder(tf.float32,shape = [None,z_dim],name = "latent_var")

    G = Generator(z,is_training = True,reuse = False)
    D_fake_logits = Discriminator(G,is_training = True,reuse = False)
    D_true_logits = Discriminator(x_img,is_training = True,reuse = True)

    #定义生成器的损失函数G_loss
    G_loss = tf.reduce_mean(tf.nn.sigmoid_cross_entropy_with_logits(
        logits = D_fake_logits,labels = tf.ones_like(D_fake_logits)))

    #定义判别器的损失函数D_loss
    D_loss_1 = tf.reduce_mean(tf.nn.sigmoid_cross_entropy_with_logits(
```

```
            logits = D_true_logits,labels = tf.ones_like(D_true_logits)))
        D_loss_2 = tf.reduce_mean(tf.nn.sigmoid_cross_entropy_with_logits(
            logits = D_fake_logits,labels = tf.zeros_like(D_fake_logits)))
        D_loss = D_loss_1 + D_loss_2

        #定义方差
        total_vars = tf.trainable_variables()
        d_vars = [var for var in total_vars if "d_" in var.name]
        g_vars = [var for var in total_vars if "g_" in var.name]

        #定义优化方式
        with tf.control_dependencies(tf.get_collection(tf.GraphKeys.UPDATE_OPS)):
            g_optimization = tf.train.AdamOptimizer(learning_rate = learning_rate,beta1 = beta1).minimize(G_loss,var_list = g_vars)
            d_optimization = tf.train.AdamOptimizer(learning_rate = learning_rate,beta1 = beta1).minimize(D_loss,var_list = d_vars)
        print("we successfully make the network")
        #网络模型构建结束

        #训练模型初始化
        start_time = time.time()          #计时
        sess = tf.Session()
        sess.run(tf.initialize_all_variables())

        #逐个epoch训练
        for i in range(epoch):
            total_batch = int(len(data)/batch_size)
            d_value = 0
            g_value = 0
            #逐个batch训练
            for j in range(total_batch):
                batch_xs = data[j* batch_size:j* batch_size + batch_size]

                #训练判别器
                z_sampled1 = np.random.uniform(low = -1.0,high = 1.0,size = [batch_size,z_dim])
                Op_d,d_ = sess.run([d_optimization,D_loss],feed_dict = {x: batch_xs,z: z_sampled1})

                #训练生成器
                z_sampled2 = np.random.uniform(low = -1.0,high = 1.0,size = [batch_size,z_dim])
                Op_g,g_ = sess.run([g_optimization,G_loss],feed_dict = {x: batch_xs,z: z_sampled2})

                #尝试生成影像并保存
                images_generated = sess.run(G,feed_dict = {z: z_sampled2})
                d_value + = d_/total_batch
                g_value + = g_/total_batch
                plot_and_save(i,images_generated)
```

```
#输出时间和损失函数loss
hour = int((time.time() - start_time)/3600)
min = int(((time.time() - start_time) - 3600 * hour)/60)
sec = int((time.time() - start_time) - 3600 * hour - 60 * min)
print("Time: ",hour,"h",min,"min",sec,"sec,"  Epoch: ",
    i,"G_loss: ",g_value,"D_loss: ",d_value)
```

开始训练：

```
if __name__ == "__main__":
    training()
```

训练时设置 epoch = 500,训练的效果如图 8.6 所示。

任务总结：从模型在训练过程中保存的生成图片样例中可以发现大部分图片主体明确,色彩逼真,图片多样性较丰富,图片效果较为贴近数据集中真实的图片。同时也能发现仍有少量生成图片损坏,无法通过人眼辨识图片主体。本项目是图片生成的一个具体实践,通过生成对抗网络的建模能力,可以对人脸的图片进行重构和简单生成。

图 8.6 训练效果图

小　　结

生成对抗网络的出现在一定程度上解决了由于深度神经网络在小规模数据集上难以训练,容易出现过拟合的问题,和其他模型相比,生成对抗网络既可以有效提升分类器的分类性能,同时生成的图像数据和真实数据相比具有语义的相似性和内容的多样性。由于生成对抗网络在理论方面较新颖,实现方面也有很多可以改进的地方,大大地激发了学术界的研究兴趣。生成对抗网络取得了实质性进展。

练　　习

代码实现 GAN 算法生成手写体数字识别数据集 MNIST 生成任务。

单元9 深度学习模型优化

深度学习作为现今机器学习领域中的重要技术手段,在图像识别领域的应用已经很成熟,并获得了很好的成果。在使用深度学习模型时,即使在数据集和模型结构完全相同的情况下,选择不同的优化算法可能导致截然不同的训练效果,甚至相同的优化算法,但是选择了不同的参数初始化策略或者训练策略,也可能会导致不同的训练结果。因此有必要学习并掌握常用的深度学习模型优化方法。

通过前面单元的学习,相信读者已经掌握了神经网络的工作原理,并且能够完成各类复杂网络的设计。在设计好模型的基础上,要想获得更高的任务指标,需要不断提高模型的学习能力,需要优化模型结构,需要提升模型的通用性,这就是模型优化的过程。在本单元中,我们将学习如何对设计好的模型进行优化,并掌握常用的模型优化技巧。本单元的知识导图如图9.1所示。

图9.1 知识导图

课程安排

课程任务	课程目标	安排课时
掌握常用深度学习模型优化方法	掌握深度学习模型优化思路,掌握常用模型优化方法,并通过实践加深理解	3

9.1 模型优化思路

模型优化本身是个很宽泛的问题，也是目前学界一直探索的目的，而从目前常规的手段上来说，一般可取如下几点：

1. 数据角度

增强数据集。无论是有监督学习还是无监督学习，数据永远是最重要的驱动力。更多的类型数据对良好的模型能带来更好的稳定性和对未知数据的可预见性。对模型来说，"看到过的总比没看到的更具有判别的信心"。但增大数据并不是盲目的，模型容限能力不高的情况下即使增大数据也对模型毫无意义。而从数据获取的成本角度，对现有数据进行有效的扩充也是个非常有效且实际的方式。良好的数据处理，常见的处理方式如数据缩放、归一化和标准化等。

2. 模型角度

模型的容限能力决定着模型可优化的空间。在数据量充足的前提下，对同类型的模型，增大模型规模来提升容限无疑是最直接和有效的手段。但越大的参数模型优化也会越难，所以需要在合理的范围内对模型进行参数规模的修改。而不同类型的模型，在不同数据上的优化成本都可能不一样，所以在探索模型时需要尽可能挑选优化简单、训练效率更高的模型进行训练。

3. 调参优化角度

如果你知道模型的性能为什么不再提高了，那已经向提升性能跨出了一大步。超参数调整本身是一个比较大的问题。一般可以包含模型初始化的配置，优化算法的选取、学习率的策略以及如何配置正则和损失函数等。这里需要提出的是对于同一优化算法，相近参数规模的前提下，不同类型的模型总能表现出不同的性能。这实际上就是模型优化成本。从这个角度的反方向来考虑，同一模型也总能找到一种比较适合的优化算法。所以确定了模型后选择一个适合模型的优化算法也是非常重要的手段。

4. 训练角度

很多时候我们会把优化和训练放在一起。但这里分开来讲，主要是为了强调充分的训练。在越大规模的数据集或者模型上，诚然一个好的优化算法总能加速收敛。但在未探索到模型的上限之前，永远不知道训练多久算训练完成。所以在改善模型上充分训练永远是最必要的过程。充分训练的含义不仅仅只是增大训练轮数。有效的学习率衰减和正则同样是充分训练中非常必要的手段。

9.2 参数初始化

在深度学习的模型中，从零开始训练时，权重和偏置的初始化有时候会对模型训练产生较大的影响。良好的初始化能让模型快速、有效地收敛，而糟糕的初始化会使得模型无法训练。

目前，大部分深度学习框架都提供了各类参数初始化方式，其中常用的有以下几种。

1. 常数初始化

把权值或者偏置初始化为一个常数。例如设置为0,偏置初始化为0较为常见,权重很少会初始化为0。TensorFlow 中也有 zeros_initializer、ones_initializer 等特殊常数初始化函数。

2. 高斯初始化

给定一组均值和标准差,随机初始化的参数会满足给定均值和标准差的高斯分布。高斯初始化是很常用的初始化方式。特殊地,在 TensorFlow 中还有一种截断高斯分布初始化(Truncated Normal Initializer),其主要为了将超过两个标准差的随机数重新随机,使得随机数更稳定。

3. 均匀分布初始化

给定最大最小的上下限,参数会在该范围内以均匀分布方式进行初始化,常用上下限为(0,1)。

4. xavier 初始化

要训练较深的网络,防止梯度弥散,需要依赖非常好的初始化方式。xavier 就是一种比较优秀的初始化方式,也是目前最常用的初始化方式之一。其目的是使得模型各层的激活值和梯度在传播过程中的方差保持一致。本质上 xavier 还是属于均匀分布初始化,但与上述的均匀分布初始化有所不同,xavier 的上下限将在如下范围内进行均匀分布采样:

$$\left[-\sqrt{\frac{6}{n+m}}, \sqrt{\frac{6}{n+m}}\right]$$

式中,n 为所在层的输入维度,m 为所在层的输出维度。

5. kaiming 初始化(msra 初始化)

kaiming 初始化又称 msra 初始化。kaiming 初始化和 xavier 一样都是为了防止梯度弥散而使用的初始化方式。kaiming 初始化的出现是因为 xavier 存在一个不成立的假设。xavier 在推导中假设激活函数都是线性的,而在深度学习中常用的 ReLu 等都是非线性的激活函数。而 kaiming 初始化本质上是高斯分布初始化,与上述高斯分布初始化有所不同,其是个满足均值为0,方差为 $2/n$ 的高斯分布:

$$\left[0, \sqrt{\frac{2}{n}}\right]$$

式中,n 为所在层的输入维度。

9.3 学习率设置

在训练神经网络时,需要设置学习率(Learning Rate)控制参数更新的速度。学习率决定了参数每次更新的幅度,如果幅度过大,那么可能导致参数在极优值的两侧来回移动;如果幅度过小,虽然能保证收敛性,但是会大大降低优化速度,需要更多轮的迭代才能达到一个比较理想的优化效果。下面举例说明上述两种情况,假设需要最小化函数 $y = x^2$,选择初始点 $x_0 = 5$。两种情况分别如下:

1. 学习率过大(学习率为1,x 在 5 ~ -5 之间震荡)

```
import tensorflow as tf

TRAINING_STEPS = 10
LEARNING_RATE = 1

x = tf.Variable(tf.constant(5,dtype = tf.float32),name = "x")
y = tf.square(x)

train_op = tf.train.GradientDescentOptimizer(LEARNING_RATE).minimize(y)

with tf.Session() as sess:
    sess.run(tf.global_variables_initializer())
    for i in range(TRAINING_STEPS):
        sess.run(train_op)
        x_value = sess.run(x)
        print("After % s iteration(s): x% s is % f."% (i +1,i +1,x_value) )
```

运行结果如下:

```
After1 iteration(s): x1 is -5.000000.
After 2 iteration(s): x2 is 5.000000.
After 3 iteration(s): x3 is -5.000000.
After 4 iteration(s): x4 is 5.000000.
After 5 iteration(s): x5 is -5.000000.
After 6 iteration(s): x6 is 5.000000.
After 7 iteration(s): x7 is -5.000000.
After 8 iteration(s): x8 is 5.000000.
After 9 iteration(s): x9 is -5.000000.
After 10 iteration(s): x10 is 5.000000.
```

2. 学习率过小(学习率为0.001,下降速度过慢,在 901 轮时才收敛到 0.823355)

```
TRAINING_STEPS = 1000
LEARNING_RATE = 0.001

x = tf.Variable(tf.constant(5,dtype = tf.float32),name = "x")
y = tf.square(x)

train_op = tf.train.GradientDescentOptimizer(LEARNING_RATE).minimize(y)

with tf.Session() as sess:
    sess.run(tf.global_variables_initializer())
        for i in range(TRAINING_STEPS):
        sess.run(train_op)
        if i % 100 = = 0:
            x_value = sess.run(x)
            print("After % s iteration(s): x% s is % f."% (i +1,i +1,x_value))
```

运行结果如下：

```
After1 iteration(s): x1 is 4.990000.
After 101 iteration(s): x101 is 4.084646.
After 201 iteration(s): x201 is 3.343555.
After 301 iteration(s): x301 is 2.736923.
After 401 iteration(s): x401 is 2.240355.
After 501 iteration(s): x501 is 1.833880.
After 601 iteration(s): x601 is 1.501153.
After 701 iteration(s): x701 is 1.228794.
After 801 iteration(s): x801 is 1.005850.
After 901 iteration(s): x901 is 0.823355.
```

TensorFlow 提供了一种更加灵活的学习率设置方法——指数衰减。通过指数衰减，可以先使用较大的学习率得到一个比较优的解，然后随着迭代逐步减小学习率，使模型在训练后期更加稳定。

9.4 优化算法选择

梯度下降法（Gradient Descent）及其一些变种算法是目前深度学习中最常用于求解凸优化问题的优化算法。神经网络很可能存在很多局部最优解，而非全局最优解。为了防止陷入局部最优，通常会采用如下一些方法，当然，这并不能保证一定能找到全局最优解，或许能得到一个比目前更优的局部最优解也是不错的。

1. Stochastic Gradient Descent/Mini-Batch Gradient Descent

在梯度下降算法中，每次的梯度都是从所有样本中累计获取的，这种情况最容易导致梯度方向过于稳定一致，且更新次数过少，容易陷入局部最优。而随机梯度下降（Stochastic Gradient Descent，SGD）是随机梯度下降的另一种极端更新方式，其每次都只使用一个样本进行参数更新，这样更新次数大大增加也就不容易陷入局部最优。但引出的一个问题在于其更新方向过多，导致不易于进一步优化。Mini-Batch Gradient Descent 便是两种极端的折中，即每次更新使用一小批样本进行参数更新。Mini-Batch Gradient Descent 是目前最常用的优化算法。

2. 动量

动量（Momentum）也是梯度下降中常用的方式之一，随机梯度下降的更新方式虽然有效，但每次只依赖于当前批样本的梯度方向，这样的梯度方向依然很可能是随机的。动量就是用来减少随机，增加稳定性。其思想是模仿物理学的动量方式，每次更新前加入部分上一次的梯度量，这样整个梯度方向就不容易过于随机。一些常见情况时，如上次梯度过大，导致进入局部最小点时，下一次更新能很容易借助上次的大梯度跳出局部最小点。

3. 自适应学习率

无论是梯度下降还是动量，优化角度都是梯度方向。而学习率则是用来直接控制梯度更新幅度的超参数。自适应学习率的优化方法很多，如 RMSprop 和 Adam。两种自适应学习率的方式稍有差异，但主要思想都是基于历史的累计梯度去计算一个当前较优的学习率。

9.5 Dropout

在深度学习模型中,如果模型的参数太多,而训练样本又太少,训练出来的模型很容易产生过拟合现象。在训练神经网络时经常会遇到过拟合问题,过拟合具体表现在:模型在训练数据上损失函数较小,预测准确率较高;但是在测试数据上损失函数比较大,预测准确率较低。为了解决过拟合问题,一般会采用正则化或者模型集成的方法,即在目标函数中加入正则项或者训练多个模型进行组合。此时,训练模型就成为一个很大的问题:正则化会使模型的求解难度增加;模型集成不仅训练多个模型费时,测试多个模型也很费时。而 Dropout 可以用较为简单的形式比较有效的缓解过拟合的发生,并在一定程度上达到正则化的效果。

Dropout 运作原理:如图 9.2 所示,在一次循环中先随机选择神经层中的一些单元并将其临时隐藏,然后再进行该次循环中神经网络的训练和优化过程。在下一次循环中,又将隐藏另外一些神经元,如此直至训练结束。在训练时,每个神经单元以概率 p 被保留(Dropout 丢弃率为 $1-p$);在测试阶段,每个神经单元都是存在的,权重参数 w 要乘以 p,成为 pw。

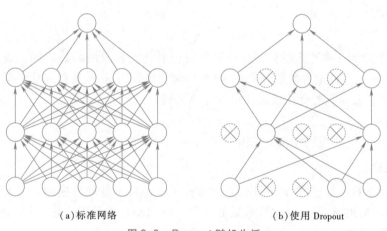

(a)标准网络　　　　　　　　　(b)使用 Dropout

图 9.2　Dropout 随机失活

9.6 批量归一化

训练深层神经网络是十分困难的,特别是在较短时间内使它们收敛更加棘手。批量归一化(Batch Normalization)是一种流行且有效的技术,可持续加速深层网络的收敛速度。结合 ResNet 中介绍的残差块,批量归一化使得研究人员能够训练 100 层以上的网络。

下面回顾一下训练神经网络时出现的一些实际挑战。

首先,数据预处理的方式通常会对最终结果产生巨大影响。使用真实数据时,第一步是标准化输入特征,使其平均值为 0,方差为 1。直观地说,这种标准化可以很好地与优化器配合使用,因为它可以将参数的量级进行统一。

其次,对于典型的卷积神经网络,训练时中间层中的变量可能具有更广的变化范围:

不论是沿着从输入到输出的层,跨同一层中的单元,或是随着时间的推移,模型参数随着训练更新变幻莫测。批量归一化的发明者非正式地假设,这些变量分布中的这种偏移可能会阻碍网络的收敛。直观地说,如果一个层的可变值是另一层的 100 倍,这可能需要对学习率进行补偿调整。

第三,更深层的网络很复杂,容易过拟合。这意味着正则化变得更加重要。

将批量归一化应用于单个可选层(也可以应用到所有层),其原理如下:在每次训练迭代中,首先基于当前小批量处理归一化输入,即通过减去其均值并除以其标准差;然后应用比例系数和比例偏移。正是由于这个基于批量统计的标准化,才有了批量归一化的名称。

如果尝试使用大小为 1 的小批量应用批量归一化,将无法学到任何东西。这是因为在减去均值之后,每个隐藏单元将为 0。所以,只有使用足够大的小批量,批量归一化这种方法才是有效且稳定的。在应用批量归一化时,批量大小的选择可能比没有批量归一化时更重要。

下面学习一下批量归一化在实践中是如何工作的。批量归一化和其他图层之间的一个关键区别是,由于批量归一化在完整的小批次上运行,因此不能像以前在引入其他层时那样忽略批处理的尺寸大小。下面分别讨论这两种情况:全连接层和卷积层,它们的批量归一化实现略有不同。

1. 全连接层

通常,将批量归一化层置于全连接层中的仿射变换和激活函数之间。设全连接层的输入为 u,权重参数和偏置参数分别为 W 和 b,激活函数为 φ,批量归一化的运算符为 BN。那么,使用批量归一化的全连接层输出的计算公式如下:

$$h = \varphi(\mathrm{BN}(Wx + b))$$

均值和方差是在应用变换的"相同"小批量上计算的。

2. 卷积层

同样,对于卷积层,可以在卷积层之后和非线性激活函数之前应用批量归一化。当卷积有多个输出通道时,需要对这些通道的"每个"输出执行批量归一化,每个通道都有自己的拉伸(scale)和偏移(shift)参数,这两个参数都是标量。

假设微批次包含 m 个示例,并且对于每个通道,卷积的输出具有高度 p 和宽度 q。那么对于卷积层,在每个输出通道的 $m \cdot p \cdot q$ 个元素上同时执行每个批量归一化。因此,在计算平均值和方差时,会收集所有空间位置的值,然后在给定通道内应用相同的均值和方差,以便在每个空间位置对值进行归一化。

3. 推理过程中的批量归一化

批量归一化在训练模式和预测模式下的行为通常不同。首先,将训练好的模型用于推理时,不再需要样本均值中的噪声以及在微批次上估计每个小批次产生的样本方差了。其次,可能需要使用模型对逐个样本进行预测。一种常用的方法是通过移动平均估算整个训练数据集的样本均值和方差,并在预测时使用它们得到确定的输出。可见,批量归一化层在训练模式和预测模式下的计算结果也是不一样的。

9.7 梯度爆炸/消失

本质上,梯度消失和爆炸是一种情况。在深层网络中,由于网络过深,如果初始得到的梯度过小,或者传播途中在某一层上过小,则在之后的层上得到的梯度会越来越小,即产生了梯度消失。梯度爆炸也是同样的。一般地,不合理的初始化以及激活函数,如 sigmoid 等,都会导致梯度过大或者过小,从而引起消失/爆炸。

下面分别从网络深度角度以及激活函数角度进行解释:

1. 网络深度

在网络很深时,若权重初始化较小,各层上相乘得到的数值都是 0~1 之间的小数,而激活函数梯度也是 0~1 之间的数。那么连乘后,结果数值就会变得非常小,导致梯度消失。若权重初始化较大,大到乘以激活函数的导数都大于 1,那么连乘后,可能会导致求导的结果很大,形成梯度爆炸。

2. 激活函数

如果激活函数选择不合适,比如使用 sigmoid,梯度消失就会很明显了,如图 9.3 所示,图 9.3(a)所示为 sigmoid 的函数图,图 9.3(b)所示为其导数的图像,如果使用 sigmoid 作为损失函数,其梯度是不可能超过 0.25 的,这样经过链式求导后,很容易发生梯度消失。

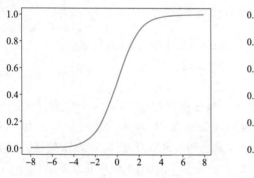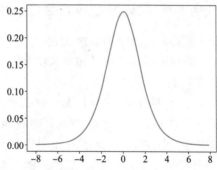

图 9.3 sigmod 函数与其导数

针对梯度爆炸和梯度消失问题,常用的解决方法有以下几类:

(1) 预训练加微调

其基本思想是每次训练一层隐节点,训练时将上一层隐节点的输出作为输入,而本层隐节点的输出作为下一层隐节点的输入,此过程就是逐层"预训练"(pre-training);在预训练完成后,再对整个网络进行"微调"(fine-tunning)。此思想相当于是先寻找局部最优,然后整合起来寻找全局最优,此方法有一定的好处,但是目前应用的不是很多。

(2) 梯度剪切、正则

梯度剪切这个方案主要是针对梯度爆炸提出的,其思想是设置一个梯度剪切阈值,然后更新梯度的时候,如果梯度超过这个阈值,那么就将其强制限制在该范围之内。这可以防止梯度爆炸。另外一种解决梯度爆炸的手段是采用权重正则化(weithts regularization),比较常见的是 L1 和 L2 正则。

(3) ReLu、leakReLu 等激活函数

①ReLu：其函数的导数在正数部分是恒等于 1，这样在深层网络中，在激活函数部分就不存在导致梯度过大或者过小的问题，缓解了梯度消失或者爆炸。同时也方便计算。当然，其也存在一些缺点，例如过滤到了负数部分，导致部分信息丢失，输出的数据分布不再以 0 为中心，改变了数据分布。

②leakReLu：就是为了解决 ReLu 的 0 区间带来的影响，其数学表达为：leakrelu = max(k∗x,0) 其中 k 是 leak 系数，一般选择 0.01 或者 0.02，或者通过学习而来。

(4) 批量归一化

批量归一化是深度学习发展以来提出的最重要的成果之一，目前已经被广泛地应用到了各大网络中，具有加速网络收敛速度、提升训练稳定性的效果，批量归一化本质上是解决反向传播过程中的梯度问题。批量归一化通过规范化操作将输出信号 x 规范化到均值为 0、方差为 1，保证网络的稳定性。

(5) 残差结构

残差的方式，能使得深层的网络梯度通过跳级连接路径直接返回到浅层部分，使得网络无论多深都能将梯度进行有效回传。

(6) LSTM

LSTM 全称是长短期记忆网络(long-short term memory networks)，是不那么容易发生梯度消失的，主要原因在于 LSTM 内部复杂的"门"，在计算时，将过程中的梯度进行了抵消。

单元任务 19　优化简化版的手写体数字识别网络

本任务将利用所学知识来优化一个简化版的手写体数字识别网络，希望同学们在实践过程中能够灵活运用所学知识来优化图像识别模型。

步骤 1：数据预处理。

本任务所使用的数据集是 MNIST 数据集。每张图片包含 28×28 个像素，把这个数组展开成一个向量是 28×28 = 784，所以在 MNIST 训练集中，输入的是一个形状为 [60000, 784] 的张量，图片中某个像素的强度值在 0~1 之间。MNIST 数据集的标签是 0~9 的数字，把它转化成 one-hot 的形式，比如：数字 0 表示为 ([1,0,0,0,0,0,0,0,0,0])；数字 5 表示为 ([0,0,0,0,0,1,0,0,0,0])。所以标签是一个 [6000,10] 的像素矩阵。

步骤 2：模型搭建与分析。

首先构建一个简单的神经网络，完成基本功能，该网络只包含输入层和输出层两个层。中间没有隐含层，输入层的神经元个数是 784，输出层的神经元个数是 10。后续会基于该基础版本进行优化。

```
import numpy as np
from tensorflow.examples.tutorials.mnist import input_data
import tensorflow as tf

#载入数据集
mnist = input_data.read_data_sets(r'./project/MNIST_data',one_hot = True)
```

```python
#设定每次训练的批次大小
batch_size = 100
n_batch = mnist.train.num_examples //batch_size

#定义输入输出占位符
x = tf.placeholder(tf.float32,[None,784])
y = tf.placeholder(tf.float32,[None,10])

#创建一个简单的神经网络
W = tf.Variable(tf.constant(0.1,shape = [784,10]))
b = tf.Variable(tf.zeros([10]))
prediction = tf.nn.softmax(tf.matmul(x,W) + b)


#定义均方误差为代价函数
loss = tf.reduce_mean(tf.square(y - prediction))
#使用梯度下降法
train_step = tf.train.GradientDescentOptimizer(0.2).minimize(loss)

#初始化参数
init = tf.global_variables_initializer()

#计算准确率
correct_prediction = tf.equal(tf.argmax(y,1),tf.argmax(prediction,1))
accuracy = tf.reduce_mean(tf.cast(correct_prediction,tf.float32))

#模型训练及测试
with tf.Session() as sess:
    sess.run(init)
    for epoch in range(201):
        for batch in range(n_batch):
            batch_xs,batch_ys = mnist.train.next_batch(batch_size)
            sess.run(train_step,feed_dict = {x:batch_xs,y:batch_ys})

        acc = sess.run(accuracy,feed_dict = {x:mnist.test.images,y:mnist.test.labels})
#测试验证
        print('epoch' + str(epoch) + ',Test accuracy' + str(acc))
```

运行结果如下：

epoch190,Test accuracy0.9211
epoch191,Test accuracy0.9223
epoch192,Test accuracy0.9220
epoch193,Test accuracy0.9219
epoch194,Test accuracy0.9227
epoch195,Test accuracy0.9232
epoch196,Test accuracy0.9228
epoch197,Test accuracy0.9235
epoch198,Test accuracy0.9220

```
epoch199,Test accuracy0.9238
epoch201,Test accuracy0.9234
```

经测试验证,准确率为 92% 左右。

步骤 3:模型优化。

下面选用几种模型优化方法提升模型的准确率,优化过程如下:

① 添加三层隐含层,其中第一层隐含层的神经元个数为 600,第二层隐含层神经元个数为 400,第二层隐含层神经元个数为 200。

② 初始化一个 0.001 学习率,随着训练次数的增加,学习率越来越小,目的是使代价函数能更好地达到最小值(为了便于理解,此处没有调用 TensorFlow 接口)。

③ 采用高斯初始化。

④ 使用 Dropout。

⑤ 使用的代价函数改为交叉熵代价函数,隐含层的激活函数需用 sigmoid,优化器改为 AdamOptimizer。

```python
import numpy as np
from tensorflow.examples.tutorials.mnist import input_data
import tensorflow as tf

#载入数据集
mnist = input_data.read_data_sets(r'./project/MNIST_data',one_hot = True)

#设定每次训练的批次大小
batch_size = 100
n_batch = mnist.train.num_examples //batch_size
#定义输入/输出占位符
x = tf.placeholder(tf.float32,[None,784])
y = tf.placeholder(tf.float32,[None,10])

#定义初始值为 0.01 的学习率
learningrate = tf.Variable(0.001,dtype = tf.float32)

#dropout 占位符
keep_prob = tf.placeholder(tf.float32)

#创建神经网络
W1 = tf.Variable(tf.truncated_normal([784,600],stddev = 0.1))
b1 = tf.Variable(tf.zeros([600]) +0.1)
L1 = tf.nn.sigmoid(tf.matmul(x,W1) +b1)
L1_drop = tf.nn.dropout(L1,keep_prob)

W2 = tf.Variable(tf.truncated_normal([600,400],stddev = 0.1))
b2 = tf.Variable(tf.zeros([400]) +0.1)
L2 = tf.nn.sigmoid(tf.matmul(L1_drop,W2) +b2)
L2_drop = tf.nn.dropout(L2,keep_prob)
```

```
W3 = tf.Variable(tf.truncated_normal([400,200],stddev = 0.1))
b3 = tf.Variable(tf.zeros([200]) + 0.1)
L3 = tf.nn.sigmoid(tf.matmul(L2_drop,W3) + b3)
L3_drop = tf.nn.dropout(L3,keep_prob)

W4 = tf.Variable(tf.truncated_normal([200,10],stddev = 0.1))
b4 = tf.Variable(tf.zeros([10]) + 0.1)
prediction = tf.nn.softmax(tf.matmul(L3_drop,W4) + b4)

#交叉熵代价函数
loss = tf.reduce_mean(tf.nn.softmax_cross_entropy_with_logits(labels = y,logits = prediction))
#使用 AdamOptimizer 优化代价函数
train_step = tf.train.AdamOptimizer(learningrate).minimize(loss)
#初始化参数
init = tf.global_variables_initializer()

#计算准确率
correct_prediction = tf.equal(tf.argmax(y,1),tf.argmax(prediction,1))
accuracy = tf.reduce_mean(tf.cast(correct_prediction,tf.float32))

with tf.Session() as sess:
    sess.run(init)
    for epoch in range(51):
        sess.run(tf.assign(learningrate,0.01*(0.95 ** epoch)))     #每次训练,学习率都减小为上一次的 0.95
        for batch in range(n_batch):
            batch_xs,batch_ys = mnist.train.next_batch(batch_size)
            sess.run(train_step,feed_dict = {x:batch_xs,y:batch_ys,keep_prob:0.5})

        acc = sess.run(accuracy,feed_dict = {x:mnist.test.images,y:mnist.test.labels,keep_prob:1.0})
        print('epoch' + str(epoch) + ',Test accuracy' + str(acc))
```

运行结果如下:

```
epoch40,Test accuracy0.9819
epoch41,Test accuracy0.9817
epoch42,Test accuracy0.9822
epoch43,Test accuracy0.9814
epoch44,Test accuracy0.9823
epoch45,Test accuracy0.9830
epoch46,Test accuracy0.9826
epoch47,Test accuracy0.9829
epoch48,Test accuracy0.9831
epoch49,Test accuracy0.9833
epoch50,Test accuracy0.9832
```

经测试验证,模型准确率为 98% 左右。

任务总结：经过实验验证发现，采取优化方法后，在提升模型训练速度的同时可以有效提升模型的准确率。本任务是一个简单的图像识别任务，所构造的网络模型也比较简单，目的是希望同学们能够通过动手实践深入理解模型优化的思路及具体操作过程，感兴趣的同学可以利用本单元所学知识优化前述几个单元的网络模型。

小　　结

本单元介绍了一些常用的深度学习模型优化技巧，希望同学们能够灵活运用所学知识来优化自己的模型。随着应用领域的深入，模型优化技巧也在不断更新。在实践中，经验丰富的算法工程师和研究人员会培养出直觉，能够灵活组合各类技巧提升模型性能。但是模型的优化方法并非一成不变，如果想在某项任务上达到最佳性能，就需要选择优化方法、重新训练模型，如此不停地重复来改进模型。目前很多深度学习框架已经开始提供自动优化工具，虽然目前这种技术仍处于初始阶段，但已经能够完成一些基本的优化功能。

练　　习

练习1　选用不同的参数初始化方法，比较分析模型运行结果。

练习2　选用不同的优化算法，比较分析模型运行结果。

练习3　设置不同的初始学习率，比较分析模型运行结果。

单元10 深度学习模型部署

模型部署是将保存和筛选后的模型转化为特殊的格式后应用到生产生活实践中的过程，我们不仅要会使用深度学习框架设计开发智能的模型，而且要结合用户实际使用的设备和应用提供智能便捷的服务，模型部署在人工智能项目开发流程中是至关重要的一环。TensorFlow 框架具有完整的工业部署生态，对于智能移动终端、边缘计算设备和嵌入式 IoT 设备，TensorFlow lite 具有强大的推理部署优势；对于浏览器前端的应用程序，TensorFlow js 可以在网页应用中灵活部署应用；对于大数据分布服务器项目，TensorFlow serving 提供了稳定且高效的部署方式。整体上模型的部署具有高度的灵活性，在模型部署过程中更能体现工程师对项目开发的整体理解。

本单元主要通过边缘端部署、浏览器前端部署和服务器部署的各个不同的生产实践环境向读者深入浅出地介绍深度学习模型部署的方法和过程。本单元的知识导图如图 10.1 所示。

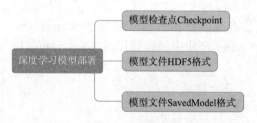

图 10.1 知识导图

课程安排

课程任务	课程目标	安排课时
模型文件简介	深入了解 TensorFlow 框架保存的三种模型文件和相应文件的具体含义，基本掌握各种模型文件的保存和重载方法	1
使用 TensorFlow lite 部署模型	熟练掌握 Python 版的模型转换和模型部署程序，了解其他移动终端模型部署的程序	4
使用 TensorFlow js 部署模型	了解 TensorFlow js 在浏览器前端的部署方法	2
使用 TensorFlow serving 部署模型	了解 TensorFlow serving 针对服务器的部署方法，熟悉 Flask 框架内与 TensorFlow serving 服务的结合应用	2

TensorFlow 以及内置的 keras 保存的模型文件主要有三种类型，Checkpoint 格式、HDF5 格式和 SavedModel 格式，下面分别介绍这三种格式的模型文件在 TensorFlow2 中的模型保存和模型重载。

10.1 模型检查点 Checkpoint

模型的检查点是模型在训练过程中的网络变量和参数状态的"快照"，模型训练是一个网络参数更新的动态过程，将该过程中某一时刻的网络模型状态记录下来，则最常用的方法就是保存为检查点 Checkpoint，这个检查点 Checkpoint 文件中则包含了模型的网络和各个参数的状态。

```
checkpoint
├─checkpoint
├─ckpt.data-00000-of-00001
├─ckpt.index
└─ckpt.meta
```

完整的 Checkpoint 检查点由四种类型的子文件构成。分别为 checkpoint 文件、ckpt.data 文件、ckpt.index 文件和 ckpt.meta 文件。

- checkpoint 文件：记录最新的检查点路径映射，用以其他函数。
- ckpt.data 文件：保存模型所有变量值、超参数值和权重值。
- ckpt.index 文件：保存模型所有变量名、超参数名和权重索引。
- ckpt.meta 文件：保存模型计算图。

在 TensorFlow1 中，检查点通过 tf.train.Saver 类的 saver 方法保存，通过加载计算图和 restore 方法重载模型。

在 TensorFlow2 中，检查点保存和重载可以通过 tf.train.Checkpoint 类的 save 方法和 restore 方法轻松实现，在 keras 模块中经常使用 tf.keras.callbacks.ModelCheckpoint 类配置保存模型状态为检查点。

```
#tf.__version__=1.x
#模型 Checkpoint 保存
saver=tf.train.Saver(max_to_keep)
saver.saver(sess,'./checkpoint',global_step=epoch)
#模型 Checkpoint 重载
saver_reload=tf.train.import_meta_graph('./checkpoint/ckpt.meta')
saver_reload.restore(sess,tf.train.latest_checkpoint('./checkpoint'))

#tf.__version__=2.x
#keras 模型 Checkpoint 保存
tf.keras.callbacks.ModelCheckpoint(filepath,save_best_only=False,save_weights_only=False,save_freq='epoch')
#模型 Checkpoint 保存
model=tf.keras.Model(...)
ckpt=tf.train.Checkpoint(model)
ckpt.save('./checkpoint')
#模型 Checkpoint 重载
ckpt=tf.train.Checkpoint()
ckpt.restore('./checkpoint').assert_consumed()
```

10.2　模型文件 HDF5 格式

HDF5 格式的模型文件是 keras 模块推荐的保存格式,可以同时将模型结构、模型权重和优化器配置保存为一个 .h5 文件。HDF5 格式的文件可以通过 save 方法保存、load_model 方法重载。

```
model = tf.keras.Sequential()
#保存完整模型
model.save('model.h5')
#仅保存模型权重
model.save_weights('model_weights.h5')
#重载完整模型
model = tf.keras.models.load_model('model.h5')
#重载模型权重
model = tf.keras.Sequential()
model.load_weights('model_weights.h5')
```

10.3　模型文件 SavedModel 格式

SavedModel 格式是 TensorFlow 用来实现跨语言、跨平台的模型文件,在 SavedModel 格式模型的保存目录下有三个文件,分别为 assets 目录、saved_model.pb 文件和 variables 目录,variables 目录下包含有 variables.data 文件和 variables.index 文件。

```
saved_model
├─assets
├─saved_model.pb
└─variables
    ├─variables.data-00000-of-00001
    └─variables.index
```

- assets 目录:资源文件目录。
- saved_model.pb 文件:保存模型为计算图,pb(protocol buffer)格式文件。
- variables 目录:ckpt 文件集合。
- variables.data 文件:保存模型所有变量值、超参数值和权重值。
- variables.index 文件:保存模型所有变量名、超参数名和权重索引。

SavedModel 格式的模型文件可以通过 tf.keras.models.save_model 方法保存模型,通过 tf.keras.models.load_model 方法重载模型。

```
#保存模型 SavedModel 格式
tf.keras.models.save_model(model,'./saved_model')
#重载模型 SavedModel 格式
loaded_model = tf.keras.models.load_model('./saved_model')
```

单元任务 20　使用 TensorFlow lite 部署模型

步骤 1:模型转换。

①准备模型。

既然要完成边缘端的部署,那么首先必须要有已经完成训练的模型,并且在本机上可以使用模型做预测,这样证明模型的有效性后再进一步部署到边缘端的设备上。本任务选用 tf2.3 环境,模型选用 ImageNet 训练的 MobileNet_v2 模型,保存模型格式为 HDF5,在命令行中输入相应的命令,从数据资源服务器下载完整的模型文件,运行结果如图 10.2 所示。

```
wget http://172.16.33.72/dataset/keras/models/mobilenet_v2_weights_tf_dim_ordering_tf_kernels_1.0_224.h5 -P ~/.keras/models
wget http://172.16.33.72/dataset/keras/models/imagenet_class_index.json -P ~/.keras/models
```

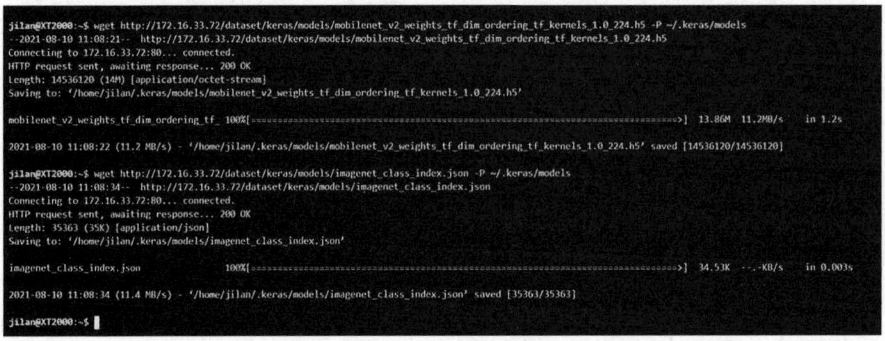

图 10.2 准备模型

准备一张待测图片,这里仍然从数据资源服务器获取,如图 10.3 所示。

```
wget http://172.16.33.72/dataset/demo/u10/image_daisy.jpg
```

图 10.3 待测图片

②在开发机上使用模型完成预测。

下面需要重载模型和相应的类别映射,展示 MobileNetV2 模型的具体架构,并使用该模型对待测的图片做图像分类的预测。在工作环境中创建预测程序 detect.py 文件,写入模型重载和预测代码。

```python
#导入模块
import os
os.environ['CUDA_VISIBLE_DEVICES'] = '0'
os.environ['TF_CPP_MIN_LOG_LEVEL'] = '3'
os.environ['TF_FORCE_GPU_ALLOW_GROWTH'] = 'true'
import tensorflow as tf
from PIL import Image

#载入待测图片
image_path = './image_daisy.jpg'
image = Image.open(image_path)
#重载模型并汇总模型
model = tf.keras.applications.mobilenet_v2.MobileNetV2()
model.summary()
#待测图片处理并预测
image = tf.image.resize(image,(224,224))
image = tf.expand_dims(image,axis = 0)
image = tf.keras.applications.mobilenet_v2.preprocess_input(image)
preds = model.predict(image)
preds_name = tf.keras.applications.mobilenet_v2.decode_predictions(preds = preds,top = 1)
print("图片预测结果:{}\t 置信度:{:.2%}".format(preds_name[0][0][1],preds_name[0][0][2]))
```

在终端输入 python detect.py 运行程序后,得到输出模型汇总和预测结果,如图 10.4 和图 10.5 所示。从结果中可以看到模型对于输入的图片的预测为 daisy(雏菊),预测的置信度为 82.06%,也就是说明模型有 82.06% 的把握预测该测试图片的类别是雏菊。

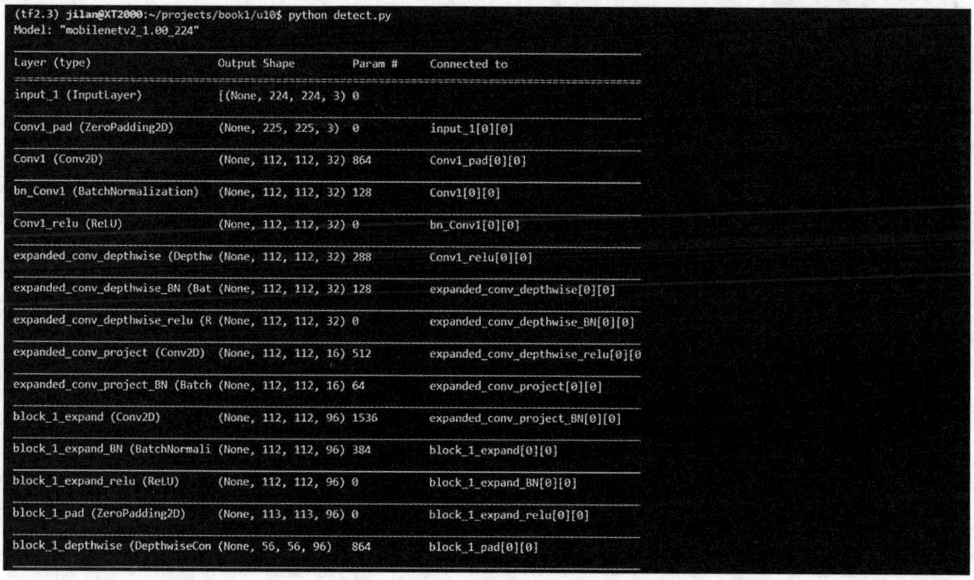

图 10.4 预测结果(一)

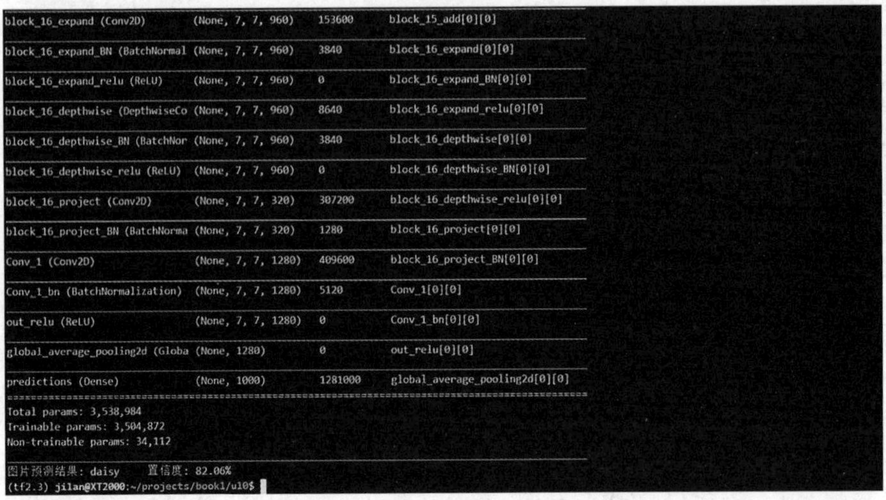

图 10.5　预测结果(二)

③TensorFlow lite 模型转换。

TensorFlow lite 是 TensorFlow 专门为智能移动终端、边缘计算和嵌入式设备开发的子框架,主要包括 TensorFlow lite 解释器和 TensorFlow lite 转换器两个主要组件。TensorFlow lite 解释器是用于在边缘端设备上运行优化后的模型,TensorFlow lite 转换器主要用来将 TensorFlow 训练好的模型转换为方便解释器推理使用的格式。用于将 TensorFlow 模型转换为 TensorFlow lite 的 API 为 tf.lite.TFLiteConverter,其中包含三种方法:

- from_keras_model:用于转换 Keas 模型。
- from_saved_model:用于转换 SavedModel 模型。
- from_concrete_functions:用于转换 Function 函数。

接下来创建模型转换程序 export_tflite.py 文件,将之前的 MobileNetV2 模型转换为 TensorFlow lite 格式。

```
import os
os.environ['CUDA_VISIBLE_DEVICES'] = '0'
os.environ['TF_CPP_MIN_LOG_LEVEL'] = '3'
os.environ['TF_FORCE_GPU_ALLOW_GROWTH'] = 'true'
import tensorflow as tf

#重载模型
model = tf.keras.applications.mobilenet_v2.MobileNetV2()

#使用 TFLiteConverter 转换模型
converter = tf.lite.TFLiteConverter.from_keras_model(model)
tflite_model = converter.convert()

#保存 model.tflite
with open("model.tflite","wb") as f:
    f.write(tflite_model)
print("模型转换完成!")
```

运行转换程序后,可以得到转换后的 model.tflite 模型文件,此时转换的模型大小约为 13.6 MB。

步骤2:模型优化。

移动边缘设备的计算能力有限,受内存和芯片硬件大小的限制,有必要进一步优化 TensorFlow lite 模型,让模型能够更好地兼容这些小型设备。这里所说的模型优化主要就是模型的量化。模型训练时通常使用 Float32 的数据类型,转换后的模型体量大、加载缓慢、计算复杂,给边缘设备的 CPU 带来严重的负载,因此我们有必要使用量化的方法进一步优化 model.tflite 模型。下面分别通过权重量化、Float16 量化和整数量化的方法对模型进行优化。

①权重量化。

权重量化的主要方式是将权重转换为 8 位精度,权重量化可以使模型尺寸减少 4 倍,适用于 CPU 加速推理。新建 weighted_tflite.py 文件,编写量化程序,运行后生成的 weighted_model.tflite 大小约为 3.5 MB。

```
import os
os.environ['CUDA_VISIBLE_DEVICES'] = '0'
os.environ['TF_CPP_MIN_LOG_LEVEL'] = '3'
os.environ['TF_FORCE_GPU_ALLOW_GROWTH'] = 'true'
import tensorflow as tf

#重载模型
model = tf.keras.applications.mobilenet_v2.MobileNetV2()

#使用 TFLiteConverter 转换模型
converter = tf.lite.TFLiteConverter.from_keras_model(model)
#权值量化
converter.optimizations = [tf.lite.Optimize.DEFAULT]
tflite_model = converter.convert()

#保存转换模型
with open("weighted_model.tflite","wb") as f:
    f.write(tflite_model)
print("模型转换完成!")
```

②Float16 量化。

Float16 量化的方式是将权重转换为 16 位浮点值,可以将模型尺度减少为原来的 1/2 倍,适用于 CPU 和 GPU 加速推理。新建 float16_tflite.py 文件,运行模型转换程序,转换后的 float16_model.tflite 大小约为 6.8 MB。

```
import os
os.environ['CUDA_VISIBLE_DEVICES'] = '0'
os.environ['TF_CPP_MIN_LOG_LEVEL'] = '3'
os.environ['TF_FORCE_GPU_ALLOW_GROWTH'] = 'true'
import tensorflow as tf
```

```
#重载模型
model = tf.keras.applications.mobilenet_v2.MobileNetV2()

#使用TFLiteConverter转换模型
converter = tf.lite.TFLiteConverter.from_keras_model(model)
#float16量化
converter.optimizations = [tf.lite.Optimize.DEFAULT]
converter.target_spec.supported_types = [tf.float16]
tflite_model = converter.convert()

#保存转换模型
with open("float16_model.tflite","wb") as f:
    f.write(tflite_model)
print("模型转换完成!")
```

③整数量化。

整数量化的主要方式是将模型所有的值转换为8位整数,通常需要提供代表性的数据集作为量化指标,整数量化可以将模型尺度减少为原来的1/3,适用于CPU和边缘TPU加速推理。新建integer_tflite.py文件,运行模型转换程序,转换后的integer_model.tflite大小约为4.0 MB。

```
import os
os.environ['CUDA_VISIBLE_DEVICES'] = '0'
os.environ['TF_CPP_MIN_LOG_LEVEL'] = '3'
os.environ['TF_FORCE_GPU_ALLOW_GROWTH'] = 'true'
import tensorflow as tf
from PIL import Image

def representative_dataset():
    for input_value in dataset.take(1):
        yield [input_value]

#载入代表数据集,项目中通常用训练集
image_path = './image_daisy.jpg'
image = Image.open(image_path)
image = tf.image.resize(image,(224,224))
image = tf.expand_dims(image,axis=0)
image = tf.keras.applications.mobilenet_v2.preprocess_input(image)
dataset = tf.data.Dataset.from_tensor_slices((image)).batch(1)

#重载模型
model = tf.keras.applications.mobilenet_v2.MobileNetV2()
#使用TFLiteConverter转换模型
converter = tf.lite.TFLiteConverter.from_keras_model(model)
#整数量化
converter.optimizations = [tf.lite.Optimize.DEFAULT]
#指定代表数据集
converter.representative_dataset = representative_dataset
```

```python
tflite_model = converter.convert()

#保存转换模型
with open("integer_model.tflite","wb") as f:
    f.write(tflite_model)
print("模型转换完成!")
```

步骤3:边缘端 Python 部署。

在边缘设备部署转换完成的 TensorFlow lite 模型,需要在边缘设备上配置相应的简单运行环境。对于边缘设备来说,只使用 CPU 计算单元做简单的推理预测,整个 TensorFlow 框架对于边缘设备来说过于臃肿,所以使用专门为移动设备准备的 tflite_runtime 库,tflite_runtime 库小巧轻便,专门用于对于边缘端 Python 部署的方案。除了 tflite_runtime 库外,边缘设备还需要 numpy 科学计算库和 pillow 图片处理库。下面编写边缘设备 t100 上的检测程序,新建 t100_detect.py 文件,在 t100_detect.py 中编写以下代码。

```python
from __future__ import absolute_import
from __future__ import division
from __future__ import print_function
import argparse
import time
import numpy as np
from PIL import Image
import tflite_runtime.interpreter as tflite


parser = argparse.ArgumentParser()
parser.add_argument('--image',default='./image_daisy.jpg',help='input image')
parser.add_argument('--model',default='./model.tflite',help='.tflite model path')
parser.add_argument('--label', default='./imagenet_labels.txt', help='index to labels')
args = parser.parse_args()

def load_labels(filename):
    with open(filename,'r') as f:
        return [line.strip() for line in f.readlines()]

interpreter = tflite.Interpreter(model_path=args.model)
interpreter.allocate_tensors()
input_details = interpreter.get_input_details()
output_details = interpreter.get_output_details()

#检查模型输入量化指标
height = input_details[0]['shape'][1]
width = input_details[0]['shape'][2]
img = Image.open(args.image).resize((width,height))
#增加一维
input_data = np.expand_dims(img,axis=0)
#输入转换
```

```python
input_data = (np.float32(input_data) / 127.5) -1
#输入数据
interpreter.set_tensor(input_details[0]['index'],input_data)
#进行推理
start_time = time.time()
interpreter.invoke()
stop_time = time.time()

output_data = interpreter.get_tensor(output_details[0]['index'])
results = np.squeeze(output_data)

top_k = results.argsort()[-1:][::-1]
labels = load_labels(args.label)
for i in top_k:
    print('图片预测结果:{}\t置信度:{:.3%}'.format(labels[i],float(results[i])))
print('推理时间:{:.3f}ms'.format((stop_time - start_time) * 1000))
```

现在还需要 ImageNet 中 1 000 种类别的标签名称文件,可以从数据资源下载得到。

```
wget http://172.16.33.72/dataset/demo/u10/imagenet_labels.txt
```

一个完整的部署文件目录共有如下四个文件。

```
.
├── image_daisy.jpg
├── imagenet_labels.txt
├── model.tflite
└── t100_detect.py
```

将此四个文件复制传输到 T100 数据推理机上即可使用推理机对图片样本进行图片分类的边缘端的测试,如图 10.6 所示。

图 10.6　未量化模型测试结果

图 10.6 显示的为未经量化的 model.tflite 模型测试结果,与之前数据处理服务器上的结果基本一致,下面将 weighted_model.tflite、float16_model.tflite 和 integer_model.tflite 部署到相同的 T00 推理机上,用同一张测试图片运行推理程序,结果如图 10.7 所示。

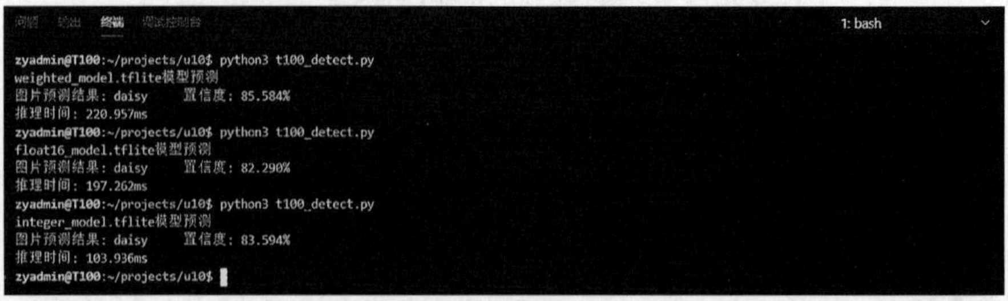

图 10.7　量化模型测试结果

从运行结果可以看出,整数量化相比权值量化、float16 量化和未量化的 tflite 模型有更为优越的推理运行速度,权值量化、float16 量化相比未量化的 TensorFlow lite 模型在单张图片预测上并没有明显的加速效果。

单元任务 21 使用 TensorFlow.js 部署模型

TensorFlow.js 是一个 JavaScript 库,在浏览器前端或 node.js 环境训练和部署机器学习模型。TensorFlow.js 的优点有:
- 不用安装驱动器和软件,通过链接即可分享智能程序。
- 网页应用,交互性强,有访问 GPS、相机端网页、手机端网页等标准 API。
- 良好安全性,数据都保存在网页客户端,不用担心数据泄露。

在浏览器前端可以通过两种方式获取 TensorFlow.js 库:

①通过脚本标签导入,添加如下脚本代码到主 HTML 文件。

```
<script src="https://cdn.jsdelivr.net/npm/@tensorflow/tfjs@3.8.0/dist/tf.min.js"></script>
```

②从 NPM 安装:npm install @tensorflow/tfjs。

步骤 1:TensorFlow.js 库使用。

在工程文件下新建一个 html 文件,导入使用 TensorFlow.js 库编写的代码。

```
<!DOCTYPEhtml>
<html>
<head>
    <meta charset="utf-8">
<title>TensorFlow.js在前端的使用</title>
    <!-- 导入 TensorFlow.js -->
    <script src="https://cdn.jsdelivr.net/npm/@tensorflow/tfjs@3.8.0/dist/tf.min.js"></script>
    <!-- 使用 TensorFlow.js 做简单的 Tensor 计算 -->
    <script>
const shape=[2,3];           //定义张量的形状
const a=tf.tensor([1,2,3,4,5,6],shape);   //tf.tensor 定义一个张量
a.print();
const b=tf.tensor2d([
    [3,-1,4],
    [1,7,-2],
]);                          //tf.tensor2d定义一个二维张量
b.print();

const a_plus_b=a.add(b);
a_plus_b.print();
const tensor_reshape=a_plus_b.reshape([6]);
tensor_reshape.print();

const c=tf.scalar(2021);     //使用 tf.scalar 定义常数
c.print();
```

```
            const d = tf.zeros([3,3]);         //全0张量
            d.print();
            const e = tf.ones([3,3]);
            e.print();
         </script>
    </head>
    <body>
        <div style = "border - style: double; border - width: 8pt; border - color: mediumblue;">
            <h1>TensorFlow.js在浏览器前端的使用</h1>
            <hr>
            <h2>按【F12】键切换到开发模式,可以在前端浏览器控制台 Console 看到程序输出的结果。</h2>
        </div>
    </body>
</html>
```

使用右键菜单中的 Open with Live Server 命令启动,用谷歌浏览器打开 http://172.16.33.106:5500/tensor.html 地址,可以显示 tensor.html 页面。按【F12】键切换到开发模式,可以在前端浏览器控制台 Console 看到程序输出的结果,如图 10.8 所示。

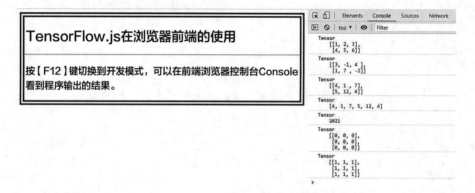

图 10.8 Tensor 计算结果

步骤 2:模型转换。

新建一个 tfjs 环境完成模型转换,在 tfjs 环境中安装 TensorFlow.js 和 flask 库。

```
pip install tensorflowjs flask flask-cors
```

新建程序 export_tfjs.py:

```
import tensorflow as tf
import tensorflowjs as tfjs

model = tf.keras.applications.mobilenet_v2.MobileNetV2()
tfjs.converters.save_keras_model(model,'./tfjs_model/')
print('模型转换完成!')
```

生成 tfjs_model 模型文件：

```
tfjs_model/
├─group1-shard1of4.bin
├─group1-shard2of4.bin
├─group1-shard3of4.bin
├─group1-shard4of4.bin
└─model.json
0 directories,5 files
```

步骤 3：配置后端服务。

要想使用 TensorFlow.js 转换的模型，还需要将 flask 服务配置为允许跨资源共享（CORS），这样 JavaScript 程序就可以正常提取文件。新建 run.py 文件，将 URL 路径映射到模型文件的 tfjs_model 目录，则模型文件可以通过 {host}/model 访问。

```python
from flask import Flask
from gevent import pywsgi
from flask_cors import CORS
from flask_cors import cross_origin

#Flask 服务
app = Flask(__name__,static_url_path = '/tfjs_model',static_folder = 'tfjs_model')

#配置 CORS
cors = CORS(app, resources = r'/*',supports_credentials = True)

#注册路由
@app.route('/')
def hello():
    return '''
    <!DOCTYPE html>
    <html>
    <head>
        <title>载入模型</title>
    </head>
    <body>
    <div style = "border-style:double;border-width:8pt;border-color:mediumblue;">
        <h1>模型载入成功！</h1>
        <hr>
        <h2>跨域模型域地址为 http://{IP}:8888/tfjs_model/model.json</h2>
        <p>注：IP 为数据处理服务器静态地址。</p>
    </div>
    </body>
    </html>
    '''

#开启服务
if __name__ == '__main__':
    server = pywsgi.WSGIServer(('0.0.0.0',8888),app)
    print("启动 Flask 服务！")
    server.serve_forever()
```

运行 run.py 程序,在浏览器窗口中输入数据处理服务器 IP 地址和端口访问 flask 服务,如图 10.9 所示。

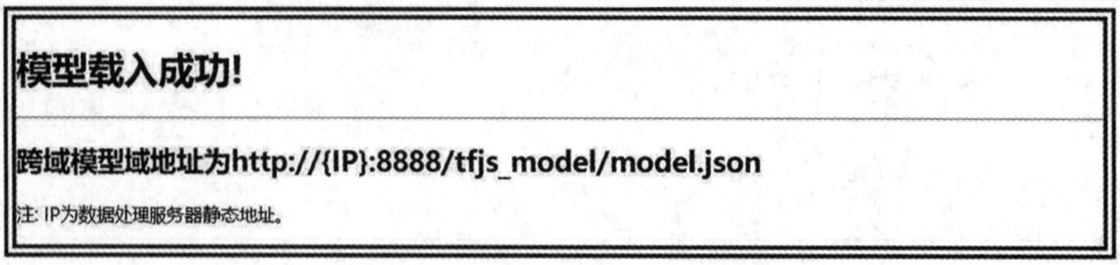

```
模型载入成功!

跨域模型域地址为http://{IP}:8888/tfjs_model/model.json

注:IP为数据处理服务器静态地址。
```

图 10.9 Flask 接口服务

步骤 4:配置前端服务。

在配置前端页面之前,首先从数据资源服务器获取一张新的待测图片 dog_Samoyed.jpg。然后新建 tfjs.html 文件,写入前端浏览器显示内容和检测图片结果。

```
#获取待测图片 dog_Samoyed.jpg
wget http://172.16.33.72/dataset/demo/u10/dog_Samoyed.jpg
```

通过 Code 编辑器在 tfjs.html 文件内写入前端代码。

```html
<!DOCTYPEhtml>

<html>
  <head>
    <title>在浏览器前端部署 TensorFlow.js,实现图片分类</title>
    <meta charset = "UTF-8">
  </head>
  <body>
    <div style = "border-style:double; border-width:8pt; border-color:mediumblue;">
      <center>
      <section>
        <h1>在浏览器前端部署 TensorFlow.js,实现图片分类</h1>
      </section>
      <section>
          <h4>检测图片</h4>
          <!-- 创建图片元素,设置 id 属性为 image -->
          <img src = "dog_Samoyed.jpg" id = "image" width = "224" height = "224"/>
          <script src = "https://cdn.jsdelivr.net/npm/@tensorflow/tfjs@3.8.0/dist/tf.min.js"></script>
          <br></br>
          <hr>
          <h4>图片分类结果</h4>
          <!-- 创建一个标题元素,用于显示预测结果 -->
          <h5 id = "output">正在预测...</h5>
          <script type = "module">
          //从 imagenet_classes.js 中获取标签列表
          import { IMAGENET_CLASSES } from './imagenet_classes.js';
          let model;
```

```
            const demo = async () = > {
              //载入模型,记得将域名替换为自己的
              model = await tf.loadLayersModel(
                'http://172.16.33.106:8888/tfjs_model/model.json'
              );
              //通过 getElementById 获取图片元素
              const imageElement = document.getElementById('image');
              //将图片元素转换为 float 格式
              const img = tf.browser.fromPixels(imageElement).toFloat();
              //将图片像素每个位置减去 127.5 再除以 127.5 以归一化到 [ -1,1 ] 之间
              const offset = tf.scalar(127.5);
              const normalized = img.sub(offset).div(offset);
              //将归一化后的图片 reshape 到模型需要的输入形状
              const batched = normalized.reshape([1,224,224,3]);
              //进行预测
              const pred = model.predict(batched);
              //获取预测结果最大值所在索引
              const index = await tf.argMax(pred,1).data();
              //从 IMAGENET_CLASSES 获取所对应的标签
              const label = IMAGENET_CLASSES[index];
              //将标签输出到 h5 元素中显示
              document.getElementById('output').innerHTML = label;
            };
            demo();
          </script>
        </section>
        </center>
      </div>
  </body>
</html>
```

在浏览器中输入数据处理服务器地址 http://{IP}:5500/tfjs.html,从前端页面可以看到前端浏览器展示的检测结果是 Samoyed(萨摩耶),如图 10.10 所示。

图 10.10　预测结果

单元任务 22 使用 TensorFlow serving 部署模型

步骤 1：转换模型。

TensorFlow serving 对 SavedModel 格式的模型具有更高的兼容性，所以在重载模型做预测时，需要使用 SavedModel 格式的模型。如果保存的模型为 HDF5 格式，那么还要进行简单的模型转换。新建模型转换程序 export_savedmodel.py，编写下述代码在 tf2.3 环境下运行，完成后即可将 HDF5 格式的 MobileNetV2 模型成功转换为 SavedModel 格式。

```
import os
os.environ['CUDA_VISIBLE_DEVICES'] = '0'
os.environ['TF_CPP_MIN_LOG_LEVEL'] = '3'
os.environ['TF_FORCE_GPU_ALLOW_GROWTH'] = 'true'
import tensorflow as tf

model = tf.keras.applications.mobilenet_v2.MobileNetV2()
model.summary()
saved_model_path = './saved_model/1'
tf.keras.models.save_model(model, saved_model_path)
print("模型转换完成!")
```

运行后生成的 saved_model 目录结构如下。

```
saved_model/1/
├── assets
├── saved_model.pb
└── variables
    ├── variables.data-00000-of-00001
    └── variables.index

2 directories, 3 files
```

同时准备一张待测图片，从数据资源服务器上获取。

```
wget http://172.16.33.72/dataset/demo/u10/dog_French_bulldog.jpg
```

步骤 2：获取 Docker 镜像。

推荐使用 Docker 安装 TensorFlow serving，这也是最便捷的安装方式。

```
docker pull tensorflow/serving
```

TensorFlow serving 可以通过两种通信协议进行服务，REST 和 gRPC。REST 通过 HTTP 发送 JSON 格式的数据实现通信，gRPC 使用 HTTP/2 协议并用 ProtoBuf 作为序列化工具实现通信。

比较项	gRPC	REST
全称	Google Remote Procedure Call	REpresentational State Transfer
传输方式	Protobuf，不可读的二进制数据	JSON，可读的数据
HTTP	HTTP/2	HTTP 1.1 / HTTP/2
性能	更快，类型安全，跨语言	跨语言
是否需要客户端	需要建立客户端	无须建立客户端
使用方法	支持任意函数	GET/PUT/POST/DELETE/....

步骤 3：TensorFlow serving 部署。

使用 Docker 启动 TensorFlow serving 服务，并同时开启 REST 与 gRPC 服务端口。

```
docker run -t-d --rm -p 9500:8500 -p 9501:8501 -v "$(pwd)/saved_model/1/:/models/mobilenet/1" -e MODEL_NAME=mobilenet tensorflow/serving
```

docker run 会利用镜像启动一个容器，如图 10.11 所示，其运行参数解释如下：
- t：为容器重新分配一个伪输入终端。
- d：后台运行。
- rm：容器退出时，自动清理容器内部的文件系统。
- p：指定要映射的 IP 和端口，其中前一个为容器端的端口，后一个为宿主机的端口。
- v：将宿主机的 saved_model/模型目录挂载到容器的/models/mobilenet 目录，容器目录必须使用绝对路径。
- e：用于传递环境变量，将 mobilenet 赋值给 MODEL_NAME。

图 10.11 启动容器

将此容器放在后台持续运行，新建一个 tfserving 虚拟环境，安装以下依赖。

```
pip install numpy opencv-python requests tensorflow-serving-api
```

新建 REST.py 文件，在程序中使用 REST 方式访问 Docker 容器的 TensorFlow serving 服务，结果如图 10.12 所示。

```
import numpy as np
import json
import requests
import tensorflow as tf
import cv2

image_path = './dog_French_bulldog.jpg'
image = cv2.imread(image_path)
image = cv2.resize(image,(224,224))
image = tf.keras.applications.mobilenet_v2.preprocess_input(image)
send_data = {'instances':[image.tolist()]}
```

```python
#REST 的访问端口为 9501
recieve_data = requests.post('http://localhost:9501/v1/models/mobilenet:predict',data=json.dumps(send_data))
#预测结果返回在'predictions'中
preds = recieve_data.json()['predictions']
#解释为标签映射名称
preds = tf.keras.applications.mobilenet_v2.decode_predictions(np.array(preds),top=1)
print("图片预测结果:{}\t置信度:{:.3%}".format(preds[0][0][1],preds[0][0][2]))
```

```
图片预测结果: French_bulldog    置信度: 63.709%
(tfserving) jilan@XT2000:~/projects/book1/u10$
```

图 10.12　预测结果

运行 REST.py 可以得到图像分类的结果为 French_bulldog(法国斗牛犬),置信度为 63.709%。

新建 gRPC.py 文件,继续使用 gRPC.py 的通信方式实现 TensorFlow serving 服务。

```python
import cv2
import numpy as np
import grpc
import tensorflow as tf
from tensorflow_serving.apis import predict_pb2
from tensorflow_serving.apis import prediction_service_pb2_grpc

image_path = './dog_French_bulldog.jpg'
image = cv2.imread(image_path)
image = cv2.resize(image,(224,224))
image = np.expand_dims(image,axis=0)
image = tf.keras.applications.mobilenet_v2.preprocess_input(image)
image_tensor = tf.make_tensor_proto(image)

#gRPC 的访问端口为 9500
channel = grpc.insecure_channel('localhost:9500')
stub = prediction_service_pb2_grpc.PredictionServiceStub(channel)
request = predict_pb2.PredictRequest()
#模型名称
request.model_spec.name = 'mobilenet'
#签名的名称,可以通过 saved_model_cli 查看
request.model_spec.signature_name = 'serving_default'
#可以使用 saved_model_cli 查看模型的输出信息
request.inputs['input_1'].CopyFrom(image_tensor)
result = stub.Predict(request,5)
print(result)
```

运行 gRPG 程序,返回经过封装的数据格式,如图 10.13 所示。

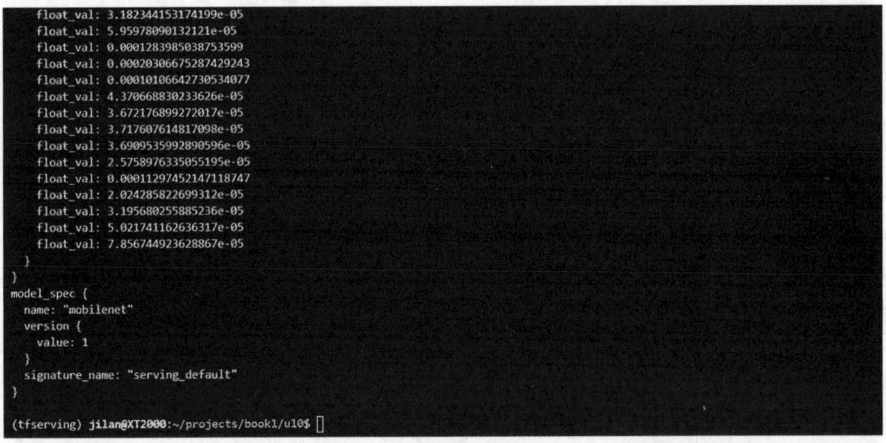

图 10.13　gRPG 返回数据

不管是 REST 通信方式还是 gRPC.py 通信方式,通常工业应用上还要再进一步做一层服务封装,如 flask 服务封装,这样就可以在更加系统的生产场景下轻松地使用 TensorFlow serving 融合成智能的应用程序,满足于各种实际的需求。

小　　结

本单元主要讲解了 TensorFlow 深度学习模型在各种生产环境下的工程部署过程。至此,已经学会了在各种场景中和不同设备上部署训练好的深度学习模型,在实际生产中需要根据具体的需求和应用场景选择不同的部署方案。

练　　习

使用 TensorFlow lite 将单元 6 中的 SSD 算法经过优化后部署在数据推理机 T100 上。

参 考 文 献

[1] REDMON J, FARHADI A. Yolov3: An incremental improvement[J]. arXiv preprint arXiv:1804.02767, 2018.
[2] LIU W, ANGUELOV D, ERHAN D, et al. Ssd: Single shot multibox detector[C]//European conference on computer vision. Springer, Cham, 2016: 21-37.
[3] RONNEBERGER O, FISCHER P, BROX T. U-net: Convolutional networks for biomedical image segmentation [C]//International Conference on Medical image computing and computer-assisted intervention. Springer, Cham, 2015: 234-241.
[4] SHUKLA N, FRICKLAS K. Machine learning with TensorFlow[M]. Shelter Island, Ny: Manning, 2018.
[5] ZACCONE G. Getting started with TensorFlow[M]. Birmingham: Packt Publishing, 2016.
[6] SANDLER M, HOWARD A, ZHU M, et al. Mobilenetv2: Inverted residuals and linear bottlenecks[C]// Proceedings of the IEEE conference on computer vision and pattern recognition. 2018: 4510-4520.
[7] HE K, ZHANG X, REN S, et al. Deep residual learning for image recognition[C]//Proceedings of the IEEE conference on computer vision and pattern recognition. 2016: 770-778.
[8] GOODFELLOW I, POUGET-ABADIE J, MIRZA M, et al. Generative adversarial nets[J]. Advances in neural information processing systems, 2014, 27.
[9] REN S, HE K, GIRSHICK R, et al. Faster r-cnn: Towards real-time object detection with region proposal networks[J]. Advances in neural information processing systems, 2015, 28: 91-99.